MW00475991

UFOS AND
THE WHITE HOUSE

UFOS AND THE WHITE HOUSE

What Did Our Presidents Know and When Did They Know It?

BY WILLIAM J. BIRNES AND JOEL MARTIN

Skyhorse Publishing

Visit our website at www.skyhorsepublishing.com.

10 9 8 7 6 5 4 3 2 1

Library of Congress Cataloging-in-Publication Data is available on file.

Cover design by Erin Seaward-Hiatt

Cover Design by Rain Saukas
Jacket Image: iStock

Print ISBN: 978-1-5107-2430-3
Ebook ISBN: 978-1-5107-2431-0

Printed in the United States of America

Dedicated to the late Jimmy Breslin from Queens Boulevard,
a columnist and reporter who always shot straight.

"Flying saucers are real, and we know what they are."
                                —President Harry Truman

CONTENTS

Introduction

OUTSIDE THE OVERTON WINDOW

Winston Churchill once noted that what we call history is actually written by the victors, describing the way the winning party line tends to prevail over the opinions of others. And so it is with American history, especially as it's taught in schools. The orthodox opinion prevails even in the face of contrary evidence. The chapters in this book will challenge the normative historical interpretation of the American presidency by presenting facts you might never encounter in a schoolroom or a college class.

Did you know, for example, that President Harry Truman, the plain-speaking "Show Me State" politician who made the decision to unleash nuclear weapons on the world, was also one of the first presidents to step forward and declare unequivocally that flying saucers were real and that the American government knew what they were? Of course you didn't.

Did you know that just days before his assassination in Dallas, President John F. Kennedy instructed the branches of the military and the CIA to release all their UFO files to the Soviets before releasing them to the American space program? Or that the president was whispering pillow talk to his mistress Marilyn Monroe about "little men from outer space" kept at a secret military location? Probably not.

How about President Reagan's not one but two UFO sightings, his administration's having been run by an astrologer, or his using an astrologer to pick his vice presidential candidate in 1980 by analyzing the star charts of his political short list? You probably knew about President Carter's UFO sighting and also that presidential candidate Hillary Clinton, the victim of 2016's Salem Witch Trials, was a fierce advocate for UFO disclosure inside the Bill Clinton White House. When you studied the history of the presidents, did you know about their statements on UFOs or the memos they wrote about them? No you didn't, and here's why.

All of these events lie outside of what journalists call the "Overton Window," the frame of acceptability for political, journalistic, and public discourse. In other words, the Overton Window is a form of censorship indicating what can be reported, what can't, and what should be relegated to what the *Huffington Post* refers to as "News of the Weird." How often will Fox's news commentators like Tucker Carlson talk about the crash of a UFO at Roswell? Rachel Maddow just about doubled over with laughter at the thought that Ronald Reagan brought up invaders from outer space at a speech before the United Nations General Assembly, but didn't report on Reagan's official UFO briefing at the White House, nor about his revelations regarding UFOs to Steven Spielberg, the director of *E.T.* and *Close Encounters of the Third Kind.*

All of these events—true, documented, and substantiated—are part of American history, but excluded from our public discourse because they lie outside the Overton Window and are unacceptable to the likes of our favorite network and cable news broadcasters. But not anymore.

Can you hear the sound of broken glass? That's the Overton Window being shattered by the chapters that follow. From the first UFO sightings by the New England colonists in the 1630s to today's White House, UFOs are making their way into American history. They're all here. And they're all true. So fasten your seat belts.

Chapter 1

GOVERNORS JOHN WINTHROP AND WILLIAM PHIPS, UFOS IN THE MASSACHUSETTS BAY COLONY, AND THE SALEM WITCH TRIALS

"In this year," wrote Massachusetts Bay Colony governor John Winthrop in his journal in 1639, "One James Everell, a sober, discreet man, and two others saw a great light in the night at Muddy River. When it stood still, it flamed up, and was about three yards square; when it ran, it was contracted into the figure of a swine: it ran as swift as an arrow towards Charlton [Charlestown], and so up and down [for] about two or three hours. They were come down in their lighter about a mile, and, when it was over, they found themselves carried quite back against the tide to the place they came from. Diverse and other credible persons saw the same light, after, about the same place."[1]

Fishermen on a small flat-bottomed cargo-carrying barge—called a "lighter"—at night on the Charles River on the edge of Boston saw a hovering elliptical-shaped object in the distance, larger than the moon, but moving up and down in the sky. The object changes shape right before their eyes, morphing into something like a flying pig. They watch this object in fascination for a couple of hours before they decide to row towards it for a closer look-see when, after traveling about a mile and suddenly and without their awareness, they find themselves back where they started. But, in order for them to have rowed back to their original starting place, they would have to have pulled against the current, which they did not do. Worse, they seem to have had no sensation of the passage of time. In today's parlance, what the fishermen experienced is called "missing time."

Perhaps modern skeptics can quibble with the reports of the description of this incident, citing the possibilities that the fishermen only saw a bright star or the planet Venus or that they might have had a few too many tots of rum before they set out on their cargo run. But Governor Winthrop attests to their sobriety in his write-up of the event and thought enough of it to include it in his official history of the Massachusetts Bay Colony. Thus, according to one of the most important political leaders of the New England colonies, about a hundred and fifty years before the signing of the Declaration of Independence, this event actually took place, belonged in the official history, and the governor vouched for the veracity of it and the boaters who recounted the event. But the story becomes more complicated, according to Governor Winthrop, because not only was the floating object in the sky witnessed by the fishermen, it was witnessed by other "credible persons," as Governor Winthrop describes them. In other words, it was a multiply witnessed aerial phenomenon of unknown origin seen and attested to by more than just the witnesses in the cargo barge. This, eliminating such things as airplanes or helicopters, because they hadn't been invented yet, counts as credible sighting as any sighting of a floating orb today.

Just five years later, towards the end of Governor Winthrop's term in office, another strange confluence of events took place, this time involving an undersea object that seemed to cause the fatal explosion of a merchant vessel belonging to Captain John Chaddock. The explosion occurred seemingly out of nowhere, and took the lives of all persons aboard Captain Chaddock's ship, stirring up rumors of a curse, of the work of the Devil, or something more mundane: a crew member who set the explosive charges because he was carrying a grudge against the vessel or its master.

If the only event had been the explosion of a wooden ship at sea, it might have passed as a simple tragedy of unknown causality. However, because the ship's explosion took place amidst the eyewitness reports of strange lights in the sky and because Chaddock's ship was reported to have hit an undersea object while at sea—well before the invention of the submarine—it is not out of the realm of possibility that Chaddock's vessel struck an unidentified submerged object, a USO. Again, witnessed by many townsfolk, strange lights began to appear in the sky that no one could explain. Were these lights demonic in origin? Was the New England colony cursed because of the townspeople's misdeeds? A sense of dread crept through the local population.

Governor Winthrop again reported in his journal that, "Exactly sixteen days after the blowing up of Capt. Chaddock's ill-fated ship and crew, and just at 'the witching hour of midnight,' as Shakespeare calls it, 'when churchyards yawn and hell itself breathes forth contagion in the air,' three men in a boat, coming toward Boston—a strange hour for reputable puritans to be out—saw two bright lights rise out of the water, at the place where the vessel had been blown up, just off the North Ferry slip. They made the still more inexplicable that the two lights assumed the form of a man, and sailed leisurely off over the water to the south, keeping but a short distance from the shore, till it reached Rowe's Wharf, where it vanished as suddenly as it had appeared just 15 minutes before."[2]

This time, instead of an unidentified flying object, the witnesses saw an unidentified submerged object, a USO that, upon breaking the surface of the water, became a flying object. If this were the only event after the shipwreck and loss of a crew, it would have been frightening enough to the local residents of the Massachusetts Bay Colony. But the sightings didn't stop. Less than a fortnight after the initial shipwreck/sighting, the twin lights appeared again in the sky and flew to the exact spot where Chaddock's ship exploded, which was where they entered the water.

The sightings over the Boston bay area still continued in the weeks following the second sighting. The next sighting witnessed by those who lived along the shore involved a single light, which some witnesses said was as large as the moon, which rose out of the water at the spot of the shipwreck and traveled over land to the present location of East Boston, where it encountered another illuminated flying object and merged with it. Witnesses watched in stunned silence as the lights joined with one another then separated, repeating this several times and all the while generating sparks and flames, until they finally morphed into a large single object, an actual disk as big as the moon, before it disappeared behind a hill overlooking Boston proper. And all of this during the tenure of Massachusetts Bay Colony Governor John Winthrop, a prudent, honest Puritan not given to flights of fancy or whimsical illusions.

While these UFO sightings—and we call them UFOs because they were truly unidentified flying objects, whatever their origin might have been—while strange were not the first UFO or USO sightings in the New World. Indeed, even before the British came to North America, even before the Puritans settled in New England, Italian explorer, sailing under commission from Ferdinand and Isabella of Spain, Christopher Columbus witnessed a self-illuminated unidentified object moving through the water alongside his flagship the *Santa Maria* as it approached the island of Hispaniola. The object seemed to be tracking the *Santa Maria* until it broke the surface of the water, rose into the sky, and flew off into the heavens. Columbus was so transfixed by the object, all he could do was stare until it was out of sight. Then he ordered his scribe to write an entry in the ship's log, thus cataloguing the first UFO / USO sighting by a European in the New World. The year was 1492.[3]

Fire and Brimstone, Cotton Mather, and the Salem Witch Trials

Our New England colonies in the seventeenth century were roiled by superstition and a belief in the immediate presence of evil spirits as well as the Devil itself. It is no mystery, therefore, that the residents of the Massachusetts Bay Colony saw the lights in the sky and the light rising out of the water, taking on a circular shape, and then flying off not as a visitation from extraterrestrials, but a sign of evil spirits. None other than New England's fire-and-brimstone preacher, Cotton Mather, reported that he had seen strange lights orbiting around the moon, a report that has lasted through five hundred years of history and eventually made it into NASA's lunar study *NASA Technical Report R-277—Lunar Events*.[4]

Mather's report of lights floating over the moon, coming over the surface from the dark side of the lunar surface, was a startling revelation from a person who was also partially responsible for shutting down the Salem Witch Trials in the town of Salem in 1692. The brutal sentences handed down to women and men accused of practicing witchcraft after sham trials—most of the victims were hanged, not burned—were based not only upon fear and upon the sanction in the Bible, which stated society should not suffer a witch or a sorceress to live, but also allowed the seizure of property from those accused. Salem Village today is now the municipality of Danvers, Massachusetts.

Chapter 2

GEORGE WASHINGTON AND THE UFO AT VALLEY FORGE

The echoes of the surprise victory Washington won over the Hessian forces celebrating Christmas 1776 outside of Trenton were only made louder by Washington's subsequent victories at the Battle of Princeton in January 1777 and the British retreat north to occupy the Atlantic coast of New Jersey. But Washington's New Jersey victories were to be soon extinguished later in 1777 by his losses at the Battle of Brandywine and then at Germantown outside of Philadelphia, which city was occupied by the British. Washington's failure to drive the British out of Philadelphia forced him to retreat to winter quarters at Valley Forge, Pennsylvania, with an army of under 15,000 troops who were demoralized after two defeats, were severely lacking supplies, had little, if any, winter gear, and were ill-shod and starving. In fact, military historians have said that Washington's army at Valley Forge was not a disciplined army at all but a ragtag cohort of volunteer farmers and tradesmen who had yet to be paid by the newly formed Continental Congress. They were at the point of mutiny.

As the terrible winter dragged on, many of Washington's soldiers, who had already missed the harvest, now wanted to go home. They feared that in their current demoralized and ill-equipped state, they were no match for the British, and would be summarily defeated, captured, and hanged for treason against the Crown. For many, their only hope was to return to their farms and claim innocence in the face of British charges should the rebellion be put down. And Washington himself was disconsolate as he watched the condition of his army decay. And in addition to the fearsome condition of his army, Washington was also at odds with the Continental Congress for their lack of support for his troops. His men had not been paid, had not been resupplied, and had not received the support an army in the field needs from the civilian government in charge of the war effort.

The Continental Army was a mixed bag of volunteers, with teenagers serving in the ranks alongside those who, today, would be considered senior citizens in their sixties. But in the mix as well were a number of foreign officers who had been trained in European military practices and sought to train Washington's troops. They were joined by a Prussian officer who called himself Baron von Steuben, who, although not a baron and not a senior officer, was, nevertheless, a skilled military tactician and a former captain in the Prussian military, who impressed Washington with his military background. Washington put him in charge of the men to train them in field maneuvers. Joining von Steuben at Valley Forge was the Marquis de Lafayette, another skilled military commander whom Washington put in charge of a regiment and then the army in Virginia, and a military engineer, Louis Duportail, who supervised the building of fortifications for the defense of Valley Forge should the British decide to attack.

As the winter wore on, even though his men were consigned to living in canvas tents which barely sheltered them from the wind-driven cold, Washington's generals such as Henry Knox were able to secure provisions from farmers in the countryside and they were able to persuade the Congress to provide more supplies for the men. As conditions improved physically, Washington, although encouraged that his troops could survive the winter, was still forlorn over what he saw as the fate of the Revolution. And that was when he had his first UFO—if that's what we want to call it—encounter.

Washington had tried to rally his staff, who seemed hopelessly demoralized, but he, too, was staring into the abyss of winter. He often ventured out from the fortifications of his encampment to pray in the woods, apart from the eyes of his staff, and to write in his journal. Then, on one day, looking up from his writing table where he was preparing dispatches for the Congress, he beheld, standing opposite his table, a glorious vision of a "singularly beautiful female." Whether this was a William Blake vision borne out of the general's own discomfiture or whether this was an actual extra terrestrial manifestation presenting a glimpse of the future, Washington was transfixed.

According to Anthony Sherman, whose story appeared in 1880, and who described himself as a member of Washington's army at Valley Forge, this is what he wrote:

"You rightly heard of Washington's going into a thicket to pray in secret for aid and comfort from God, the interposition of his Divine Providence brought us safely through the darkest days of tribulation. One day, I remember it well, when the chilly winds whistled through the leafless trees, though the sky was cloudless and the sun shone brightly, he remained in his quarters nearly all the afternoon alone. When he came out, I noticed that his face was a shade paler than usual. There seemed to be something on his mind of more than ordinary importance. Returning just after dark, he dispatched an orderly to the quarters who was presently in attendance. After a preliminary conversation of about an hour, Washington, gazing upon his companion with that strange look of dignity which he alone commanded, related the event that occurred that day."[1]

And in his own words, Washington himself wrote in his journal:

"This afternoon, as I was sitting at this table engaged in preparing a dispatch, something seemed to disturb me. Looking up, I beheld standing opposite me a singularly beautiful female. So astonished was I, for I had given strict orders not to be disturbed, that it was some moments before I found language to inquire the cause of her presence. A second, a third and even a fourth time did I repeat my question, but received no answer from my mysterious visitor except a slight raising of her eyes.

"By this time I felt strange sensations spreading through me. I would have risen but the riveted gaze of the being before me rendered volition impossible. I assayed once more to address her, but my tongue had become useless, as though it had become paralyzed.

"A new influence, mysterious, potent, irresistible, took possession of me. All I could do was to gaze steadily, vacantly at my unknown visitor. Gradually the surrounding atmosphere seemed as if it had become filled with sensations, and luminous. Everything about me seemed to rarefy, the mysterious visitor herself becoming more airy and yet more distinct to my sight than before. I now began to feel as one dying, or rather to experience the sensations which I have sometimes imagined accompany dissolution. I did not think, I did not reason, I did not move; all were alike impossible. I was only conscious of gazing fixedly, vacantly at my companion.

"Presently I heard a voice saying, 'Son of the Republic, look and learn,' while at the same time my visitor extended her arm eastwardly, I now beheld a heavy white vapor at some distance rising fold upon fold. This gradually dissipated, and I looked upon a stranger scene. Before me lay spread out in one vast plain all the countries of the world—Europe, Asia, Africa and America. I saw rolling and tossing between Europe and America the billows of the Atlantic, and between Asia and America lay the Pacific.

"'Son of the Republic,' said the same mysterious voice as before, 'look and learn.' At that moment I beheld a dark, shadowy being, like an angel, standing or rather floating in mid-air, between Europe and America. Dipping water out of the ocean in the hollow of each hand, he sprinkled some upon America with his right hand, while with his left hand he cast some on Europe. Immediately a cloud raised from these countries, and joined in mid-ocean. For a while it remained stationary, and then moved slowly westward, until it enveloped America in its murky folds. Sharp flashes of lightning gleamed through it at intervals, and I heard the smothered groans and cries of the American people.

"A second time the angel dipped water from the ocean, and sprinkled it out as before. The dark cloud was then drawn back to the ocean, in whose heaving billows in sank from view. A third time I heard the mysterious voice saying, 'Son of the Republic, look and learn,' I cast my eyes upon America and beheld villages and towns and cities springing up one after another until the whole land from the Atlantic to the Pacific was dotted with them.

"Again, I heard the mysterious voice say, 'Son of the Republic, the end of the century cometh, look and learn.' At this the dark shadowy angel turned his face southward, and from Africa I saw an ill omened specter approach our land. It flitted slowly over every town and city of the latter. The inhabitants presently set themselves in battle array against each other. As I continued looking I saw a bright angel, on whose brow rested a crown of light, on which was traced the word 'Union,' bearing the American flag which he placed between the divided nation, and said, 'Remember ye are brethren.' Instantly, the inhabitants, casting from them their weapons became friends once more, and united around the National Standard.

"And again I heard the mysterious voice saying 'Son of the Republic, look and learn.' At this the dark, shadowy angel placed a trumpet to his mouth, and blew three distinct blasts; and taking water from the ocean, he sprinkled it upon Europe, Asia and Africa. Then my eyes beheld a fearful scene: From each of these countries arose thick, black clouds that were soon joined into one. Throughout this mass there gleamed a dark red light by which I saw hordes of armed men, who, moving with the cloud, marched by land and sailed by sea to America. Our country was enveloped in this volume of cloud, and I saw these vast armies devastate the whole country and burn the villages, towns and cities that I beheld springing up. As my ears listened to the thundering of the cannon, clashing of sword, and the shouts and cries of millions in mortal combat, I heard again the mysterious voice saying, 'Son of the Republic, look and learn.' When the voice had ceased, the dark shadowy angel placed his trumpet once more to his mouth, and blew a long and fearful blast.

"Instantly a light as of a thousand suns shone down from above me, and pierced and broke into fragments the dark cloud which enveloped America. At the same moment the angel upon whose head still shone the word Union, and who bore our national flag in one hand and a sword in the other, descended from the heavens attended by legions of white spirits. These immediately joined the inhabitants of America, who I perceived were well nigh overcome, but who immediately taking courage again, closed up their broken ranks and renewed the battle.

"Again, amid the fearful noise of the conflict, I heard the mysterious voice saying, 'Son of the Republic, look and learn.' As the voice ceased, the shadowy angel for the last time dipped water from the ocean and sprinkled it upon America. Instantly the dark cloud rolled back, together with the armies it had brought, leaving the inhabitants of the land victorious!

"Then once more I beheld the villages, towns and cities springing up where I had seen them before, while the bright angel, planting the azure standard he had brought in the midst of them, cried with a loud voice: 'While the stars remain, and the heavens send down dew upon the earth, so long shall the Union last.' And taking from his brow the crown on which blazoned the word 'Union,' he placed it upon the Standard while the people, kneeling down, said, 'Amen.'

"The scene instantly began to fade and dissolve, and I at last saw nothing but the rising, curling vapor I at first beheld. This also disappearing, I found myself once more gazing upon the mysterious visitor, who, in the same voice I had heard before, said, 'Son of the Republic, what you have seen is thus interpreted: Three great perils will come upon the Republic. The most fearful is the third, but in this greatest conflict the whole world united shall not prevail against her. Let every child of the Republic learn to live for his God, his land and the Union.' With these words the vision vanished, and I started from my seat and felt that I had seen a vision wherein had been shown to me the birth, progress, and destiny of the United States."[2]

Describing a physical sensation that alien contactees and abductees have reported for over two hundred and fifty years, General Washington wrote in his journal that the figure he saw not only provided him the courage to maintain his steady leadership of his army, but the courage of faith in the knowledge that the war of the revolution would be successful. We ask, was this a vision of an extraterrestrial or was Washington, at the moment of his deepest despair over the fate of his army, actually experiencing what, over two hundred years later, would be called a remote viewing event in which he literally traveled into the future with the help of a palpable figure appearing his own vision to show him what would befall his new country and the army that was charged with wresting its control from the British Crown?

The Battle for Fort Duquesne at the Monongahela River

Alternatively, we could also explain it as a visitation from a Native American spirit, the spirit that had protected him over twenty years earlier at the Battle of Monongahela during the French and Indian War when the British sought to take Fort Duquesne. Three sides fought in this war: the British, looking to settle the land in western Pennsylvania, the French, establishing a sea route across the three rivers—the Monongahela, the Allegheny, and the Ohio—to the Great Lakes and thence to Canada, and the native tribes including the Shawnee, Huron, and Iroquois, who wanted no part of this war, but threw in with the French against the British. The Battle of Monongahela itself was an attempt by the British under General Braddock, assisted by colonial militia Lieutenant Colonel George Washington, to take the French Fort Duquesne. In this battle, Colonel Washington was a hero. After the British had crossed the Monongahela River and had begun their march towards the fort, they were confronted by a contingent of French troops and, in a quick exchange of fire, forced them into confusion. But the Native Americans who had accompanied the French knew exactly what to do. They had seen the British formation before and how it stood its ground in square formations against an enemy charging towards them. The Native Americans formed into two war parties and took cover in the thickets surrounding the British. Then, from the cover of the foliage where they took a bead on the flanks of the redcoats, they fired volley after volley into the British troops, led by Braddock, until Braddock himself was struck by a ball.

George Washington rode to Braddock's side where he comforted the general, who was unresponsive, and commandeered a wagon to remove him to safety on the other side of the river. Now dismounted and standing in the midst of enemy fire, Washington ordered his troops to form a rear guard to cover the British retreat to the far side of the bank. And it was in this rear guard action that Washington almost met a lethal fate. A Native American chief had Washington dead in his musket sights and was prepared to fire. Washington, who by the end of the career had fought in over two hundred battles, had never been wounded. Yet, at this one moment when he was fighting on the side of the British in 1755, he was almost certainly fated to be killed, when the Native American chief stayed his hand. In his sights, the Native American chief saw not only the figure of Washington, but he saw what he described as the Great Spirit protecting Washington and his men. He ordered his braves to hold their fire, thus saving Washington's life. "Thus does Fate itself often intervene to save the undoomed righteous warrior when his courage holds" (*Beowulf* [trans.WJB], 572b-573).[3]

Now, deep in the woods of Valley Forge, Washington, retreating into a moment of prayer, saw another vision, this one even more specific to what we now know about UFO sightings of floating orbs. Washington said that when he went into the woods, he saw a floating green orb hovering in front of him. He could not explain it, but in the wake of his vision of the future, he was not afraid. As he stared at the floating green object, he said he saw what he believed to be small green-toned Native Americans jumping to the ground and running into the woods. And then he saw what he assumed to be a chief, who spoke to him as if he knew him, had protected him, and told him that he was protected by the Great Spirit. Then the chief vanished. Washington looked around him. He was alone in the snow. But, to his amazement, though he had come face to face with the figure, there were no footprints in the new fallen snow, not a one. Where the orb had touched down after hovering, there was no impression in the snow either. It was if none of it had ever happened, yet Washington knew, knew at that moment, that he had been protected and was being protected and that his mission would succeed.

A renewed and buoyant Washington returned to his staff officers with the swagger of confidence. Although his officers inquired about what had seemed to change his mood, Washington was diffident, saying nothing in particular. But they could tell that something was different. Finally, the Marquis de Lafayette approached his commanding officer to inquire as to the change in his demeanor. What had made him so suddenly imbued with a sense of victory when the army itself had been staring into the face of defeat? Washington's answer was even stranger than Lafayette could have imagined. The general explained that he saw the face of an old enemy, a chief against whom he had fought at the Battle of Monongahela and who had, for some reason, spared his life. Yet it was this chief who had somehow managed to cross the entire territory of Pennsylvania, from the banks of the Monongahela to the banks of the Schuylkill, to deliver a message of hope. Then, even as Washington spoke of his amazement to Lafayette, it dawned on him, the very chief he had just seen and who had just spoken to him had only died a year earlier. Washington realized in a flash that he had speaking to a dead man. He had received a message from beyond the grave.

Had the Great Spirit sent his own messenger to give Washington hope for the future? Was the general and future president so protected by the world of the paranormal that he could not fail? Or was this an intervention by extraterrestrials, assuming the shapes and sounds of those recognizable to Washington so he would not be affrighted at the vision? Whatever it was, it certainly motivated Washington to the point he ordered his staff to create projects for the men to keep them occupied. Spring would soon arrive and soldiers occupied with preparing for battle would be more motivated than soldiers shivering in the cold and longing for the warmth of their hearths and the security of their harvest.

By the following spring, a renewed Continental Army, now replenished through the efforts of Henry Knox, disciplined by von Steuben, and trained in tactics by the great Lafayette, who would soon take command of his own army in the South, moved out of Valley Forge, crossed the New Jersey colony to reach the British army that had retreated to the east coast, and defeated them handily at the Battle of Monmouth. Thus, by 1781, Washington had turned around the fortunes of war and forced the British, who would soon muster at Yorktown, Virginia, to rely on the help from Admiral Graves, himself holed up safely with his fleet in New York harbor.

But that help never came, and Graves was court-martialed and executed. A French fleet boxed in the British at Yorktown, a contingent of French troops also confronted the British at Yorktown, Washington had led his army to Yorktown, and, with the French fleet blocking any help from the sea, the siege at Yorktown forced the British to surrender, thus effectively ending the British campaign in the North American colonies.

George Washington would go on to become our nation's first president after the ratification of the United States Constitution. His supernatural adventures would not end, though. At times of great crisis, Washington would appear again, this time as a ghost, to lead his troops at the Battle of Gettysburg. He was not only our country's first president, but the first paranormal president, at the head of a history of UFO encounters that would mark key points in American history.

Chapter 3

PRESIDENT THOMAS JEFFERSON'S UFO REPORT TO THE AMERICAN PHILOSOPHICAL SOCIETY

In January 1801, Thomas Jefferson, our nation's then vice president, who that year would become our third president (1801-1809), reported to the prestigious American Philosophical Society, of which he was also the president, the story of a UFO sighting over Baton Rouge that, according to the society's proceedings minutes, was "as large as a house." The information of the sighting came to Jefferson from a well-respected scholar of his day, William Dunbar, an astronomer who told Jefferson that the event was multiply witnessed by citizens who were sober, serious, and not given to flights of fancy. As the incoming president, Jefferson was, along with Harry Truman in 1950, the highest-ranking federal official to report on UFOs while still in office even though other future presidents, Ford, Carter, and Reagan, all either acknowledged the existence of UFOs as witnesses or sought disclosure from the government before they entered the Oval Office.

Thomas Jefferson has often been called an American Renaissance man, an internationalist, architect, landscape designer, encyclopedist in the French tradition, diplomat, statesman who crafted the inspiring words of the Declaration of Independence, and a skilled politician who served two terms as president. In addition to his other accomplishments, Jefferson also led the American Philosophical Society, the closest thing America had to a Royal Academy. As an open-minded—although not open-minded enough to grant emancipation for the over 600 slaves at his plantation he called Monticello—inquirer about the world around him, Jefferson was also open to the current thinking of his day, including the notions of French author Bernard le Bovier de Fontenelle who in 1686 wrote in *Conversations on the*

Plurality of Worlds that there was a multiplicity of worlds and that Earth was just one of many inhabited planets.[1] Jefferson was joined in this belief by another major eighteenth-century American scholar, Benjamin Franklin. George Washington, who was a Freemason, also subscribed to the belief that there were many inhabited worlds in the universe and that other life forms might be as advanced as or even more advanced than human beings.

One might think that for Puritan Christians, a belief in the multiplicity of worlds and inhabitants thereupon would be a heretical belief. On the contrary, given the nature of thinking on the Continent in the middle to late eighteenth century, this was actually conventional thinking. In Voltaire's novel *Candide*, the overriding principle is that the Creator made the universe and then simply left it alone to evolve by itself. Thus, the optimistic view that once created, the universe, and by extension, humanity, was fundamentally good meant that even if life existed elsewhere, it, too, would be good because it was all a product of the Creator.[2]

Jefferson had also been a member of the Lunar Society along with Benjamin Franklin, a group of intellectual elites who, though recognizing their ties to England, also believed that the colonies were independent polities not subject to taxation by the Crown. The beliefs of the men who framed the Declaration of Independence about a society founded upon the Athenian ideals of a democracy and the Roman ideals of a senate that would advise and ultimately affirm the decisions of the emperor actually came to fruition after the Revolutionary War ended with the Treaty of Paris. Now on the precipice of forming an independent country, the original framers then turned to the thinkers of the Age of Enlightenment, who, in turn, had turned to the Greeks and Romans for their inspiration. Thomas Jefferson was a part of this intellectual movement.

Philosophe and encyclopedist Bernard le Bovier de Fontenelle's *Conversations on the Plurality of Worlds* (*Entretiens sur la pluralité des mondes*) was the inspirational treatise for the American thinkers and became one of the most talked about works on both sides of the Atlantic, forcing philosophers to consider the plausibility that humans weren't the only sentient life form in the universe. The book was way ahead of its time for a variety of reasons, all of which captured the imaginations of Jefferson and Franklin. First of all, it was based on a Copernican vision of the universe in which Earth's position as just one of a number of planets reduced it from being the absolute center of creation. It was among other worlds that revolved around the sun, and if life was created on Earth, then there was no reason the Deity did not create life on other worlds. Second, the book was almost Socratic in its conceit, comprised of conversations between a philosopher and his hostess, a marquise, about the possibilities of alien species navigating through outer space. Third, this was one of the first books in which a woman was the other half of the conversation, asking the philosopher about his vision of possibility of alien life in the universe. Accordingly, for thinkers in the latter halves of the seventeenth and eighteenth centuries, this was a treatise that defied traditional church dogma and laid the seeds, in part, for what would become eighteenth-century Deism.

Thomas Jefferson, like many of his contemporaries among the American intellectual and political elite, was a Deist. While ascribing to the teachings of Jesus, Jefferson believed that the spirit behind those teachings had been altered by Jesus' disciples and felt that the state should not be held responsible to promulgate the church's teachings. In fact, Jefferson articulated the proposition of the separation between church and state as the Framers of the Constitution debated the articles that would become the Bill of Rights. It was against this background of free thinking that Jefferson was open to all kinds of scientific and quasi-scientific interpretations of reality. It was from this perspective of being a "Unitarian of one" that Jefferson approached the story reported to him by William Dunbar of a UFO landing in Tennessee.

This was Jefferson's letter to the American Philosophical Society:

Description of a singular Phenomenon seen at Baton Rouge, by William Dunbar, Esq. communicated by Thomas Jefferson, President A. P. S.a

Natchez, June 30th, 1800

Read 16th January 1801.

A phenomenon was seen to pass Baton Rouge on the night of the 5th April 1800, of which the following is the best description I have been able to obtain.

It was first seen in the South West, and moved so rapidly, passing over the heads of the spectators, as to disappear in the North East in about a quarter of a minute.

It appeared to be of the size of a large house, 70 or 80 feet long.

It appeared to be about 200 yards above the surface of the earth, wholly luminous, but not emitting sparks; of a colour resembling the sun near the horizon in a cold frosty evening, which may be called a crimson red. When passing right over the heads of the spectators, the light on the surface of the earth, was little short of the effect of sun-beams, though at the same time, looking another way, the stars were visible, which appears to be a confirmation of the opinion formed of its moderate elevation. In passing, a considerable degree of heat was felt but no electric sensation. Immediately after it disappeared in the North East, a violent rushing noise was heard, as if the phenomenon was bearing down the forest before it, and in a few seconds a tremendous crash was heard similar to that of the largest piece of ordnance, causing a very sensible earthquake.

I have been informed, that search has been made in the place where the burning body fell, and that a considerable portion of the surface of the earth was found broken up, and every vegetable body burned or greatly scorched. I have not yet received answers to a number of queries I have sent on, which may perhaps bring to light more particulars.

Of note here is the relationship between Vice President Jefferson and William Dunbar, naturalist and scientist as well as a planter and merchant. Dunbar was a Scot of aristocratic parentage who emigrated to Philadelphia in 1771 but who wound up in Natchez, where he owned a plantation and became a surveyor for the Spanish Crown. In addition to his mercantile pursuits, Dunbar was also an inventor and an engineer as well as an observer of empirical scientific phenomena. He was particularly fascinated by life along the Mississippi River, and many of his observations involved his notes on local climate conditions, flora and fauna, and even the languages used by local Native American tribes. As such, though local, he was becoming a natural scientist who caught Jefferson's attention. Hence, Jefferson's submission of Dunbar's UFO observation to American Philosophical Society.

Dunbar also invented procedures for extracting oils from cotton seeds, but his great achievement was building a telescope at Union Hill near Natchez, where he made observations of the heavens and noted celestial phenomena. It was in that context that he noted and then reported to Jefferson what could be considered either a UFO crash landing, leaving trace evidence at the crash site, or the crash of another object, most likely a meteorite, that caused damage, but not catastrophic destruction. There had been reported crashes of meteorites in Europe, and Dunbar was aware of the news of these events. Accordingly, there might have been some confusion regarding the nature of the object that fell, even though Dunbar was more certain than not that it was not a meteorite because of the size of the object, which was a big as a house.[3]

According to Bill Thayer, who curated the Jefferson letter on his website:

More importantly, if the observation of the object's size is anywhere near accurate, it was not a meteorite: an object of this size, entering earth's atmosphere at a speed typical of objects falling to earth from space, would probably have left a much larger trace of itself, and would almost certainly have killed the observer and anyone else near the fall. Scientists currently gauge the size of the iron meteor that created Arizona's Meteor Crater, for example, at roughly 50 meters, only about twice the estimate reported by Dunbar.

Further confirmation that this was no meteorite seems to be given by the object's speed. Assuming more or less flat terrain (and though the vicinity of Baton Rouge is considered hilly by Louisiana standards, the State is one of the flattest in the Union and this area is at most gently rolling) and an observer whose eyes were a bit more than 1.50 meters above the ground the horizon is about 4.4 km away. The distance covered by the object within the witnesses' field of vision was thus a maximum 9 kilometers, but probably only about two-thirds of that (since they surely didn't notice it the instant it rose over their horizon, although once they saw it, they must with equal certainty have tracked it to the very end). If, then, it covered 6 to 9 km in something like 15 seconds, it was traveling at no more than 2200 km an hour. This is considerably less than the 11,000 km/h minimum impact velocity of an object free-falling to Earth from space. Furthermore, if we can trust Dunbar's witnesses on the height of the object above the ground, and as he explicitly states, directly above their heads—yet such perceptions of distance against a featureless sky are notoriously subject to error, even among trained pilots—its trajectory must have been far flatter than that of any normal meteor: it was 200 m above the ground and continued to travel at least 6 km (to the horizon, then "a few seconds") before it crashed, an angle of at most 1.9°. He speaks of it, at any rate, as on a more or less level trajectory.

The vague language and third-hand nature of the report on the debris field are regrettable, but the impact damage and the seismic event are consonant with a small meteor—this handy-dandy calculator courtesy of the Imperial College, London originally, backed up by some serious science, will be of interest to those of you wishing to input varying parameters— but also with a supersonic aerial craft of some kind. Inputting my own estimated parameters, most of them already given above,[4]

Distance from impact: 6 km
Projectile diameter: 75 feet
Projectile density: porous stone: 1500 kg/m3
    maybe a bit more if some kind of craft, i.e., a semi-hollow metal object
Impact velocity: 0.6 km/s
Impact angle: 1.9°
Target type: Sedimentary rock

the model yields a seismic effect somewhat, but not much, less than that reported by the witnesses (who were almost certainly not knocked down, else we'd been told), and a 55-decibel sound level, similarly less than that reported; more interestingly, a crater about 40 m wide and 8.7 m deep, and a transient crater depth of 11.6 meters, with the projectile landing "intact", and thus presumably lodged about 3 meters below the surface of the ground; get out your GPS and metal detectors, folks.

In a discussion online, two good points have been made: (a) at the range of speeds we're looking at here, there should have been a sonic boom in addition to the sound of any impact; and (b) Dunbar's estimate of the object's size, if it was on fire, is probably not so good: the most we can say is that it was no larger than a house. In turn, if it was a house-sized object coming in at a meteoric speed, it would have been a huge event, with no survivors for miles, flattened trees, etc. So of one thing we can be sure: if it was the size of a house, it crashed at a low speed; if it was a meteor, it was not the size of a house.[5]

In addition to Jefferson's fascination with empirical scientific observation, in his later years he became something of a mystic, particularly in his relationship with second president of the United States, John Adams, in whose administration he served. Although friends, Jefferson and Adams had a contentious relationship, particularly as it applied to the powers of the central federal government. Adams was a federalist while Jefferson is regarded as the founder of the Democratic Party and an advocate for states' rights. Adams believed in a strong federal government, but after his vice president, Jefferson, defeated him in the 1800 presidential election, Adams and Jefferson became bitter towards each other. Though Adams wanted to be cordial to his old friend, his wife, former First Lady Abigail Adams, remained hostile. In an exchange of letters between Jefferson and Adams, the two men, at odds politically, including their contentious disagreement over getting into a war with France, finally came together intellectually and spiritually, finding solace in a friendship they shared at the formation of the new republic. They stayed in such close communication that they even contemplated the dates of their respective deaths, wondering who would be the first to pass away. As it turned out, ironically if nothing else, the two men died on the exact same day, the same day as the signing of the Declaration of Independence, albeit fifty years later, on July 4, 1826. The two had weathered the Madison administration's prosecution of the war with Britain in 1812, no doubt shuddering over the burning of Washington, DC, and John Adams had lived long enough to see his son, John Quincy, ascend to the presidency.

When Adams lay on his deathbed he began wondering about his friend and former vice president Jefferson, the man who defeated him. The two men also assured each other that it was their intent to meet on the other side, to encounter one another in heaven after they had left this life. But who would leave this life first? Adams was reported to have said that he wondered about Jefferson, whom he believed would outlast him. As it turned out, however, on the very day that Adams died, Thomas Jefferson had died one hour earlier. Perhaps the two men did ultimately meet up in heaven—at least that's what they'd hoped.

Chapter 4

LINCOLN'S UFOS

One of the lesser-known aspects of Lincoln's belief system was that he believed not only in extraterrestrials but that they had been to this planet in ancient times. Like Benjamin Franklin and George Washington, Lincoln, though surely a Christian, also believed in the plurality of worlds, a theory espoused in the seventeenth century which said that if the Creator created Earth and all of the life forms thereon, and if the Creator created the entire universe, who's to say that there aren't more worlds inhabited with intelligent life forms? As a person who believed in the power of prophecy and the ability, manifested by his reliance on Nettie Colburn and other psychics, of some human beings to have contact with the spirit world, Lincoln also believed in supernatural portents that pointed the ways to future events. Thus it was that in March 1861, at the time of his inauguration to his first term, a singular UFO event over New York City portended the contentiousness of the incoming Lincoln administration, the Civil War, the Emancipation Proclamation, and the assassination of the president who held the union together.

Before the ratification of the Twentieth Amendment to the Constitution in 1933, setting January 20 as the presidential inauguration day, incoming presidents were inaugurated in March. So it was with Abraham Lincoln in 1861 at the very start of his first term in office. It was a propitious day in a propitious year as the country pulled in two different directions: a plantation slave-based economy or the industrial and internationalist economy of the northern states. The Republican platform, while stating unequivocally that it abhorred slavery and prohibited the slave trade in the United States, nevertheless pledged to uphold the Constitutional amendment of states' rights against the federal government when no federal law preempted state law. It was a precarious balance as the South, as it had for decades, gravitated to what was then called "disunion."

In New York, at the infamous location known as the "Five Points" in lower Manhattan where immigrant and Manhattan native gang members often clashed, residents living on nearby Baxter Street witnessed a flying saucer overhead. It was nighttime on a moonless March night, but the object witnesses saw was illuminated. Illuminated by what? The sky was pitch black yet the cross-shaped object that hovered over the city's rooftops seemed to be lit from the inside. Other people saw flying objects as well, not only over Manhattan, but across the river in New Jersey as well. Were these portents of the momentous years that Lincoln's term in office would mark or were these actual UFOs piloted by life forms that had already seen the future of the Lincoln presidency and the war he would fight and his ultimate end? Whatever the answer, Lincoln's was the most paranormal presidency and his ghost still lingers in the corridors of the White House while it awaits the next first family to be haunted by the apparition.

Chapter 5

PRESIDENT WILLIAM MCKINLEY AND THE TEXAS AIRSHIP MYSTERY

Although President McKinley himself might not have ever witnessed an actual UFO, folks in the area of Aurora, Texas, reported that they certainly did in April 1897, less than a month after McKinley was inaugurated for his first term. He would win a second term in the election in 1900 and die in office a few months later, a victim of the zero-year curse. But just a couple of weeks after he took office, a strange metallic craft was observed by local residents flying low over Wise and Erath counties in Texas. This lone craft's appearance wasn't the first time folks in the area would see a flying craft they could not identify. From 1896 into the spring of 1897, slow-moving flying objects were spotted not only in Texas, but in various locations in the skies over the American southwest. Although no one seemed particularly alarmed by the sightings, the newspapers were abuzz with stories and hand-drawn images of the craft, some of which were adorned with wings to look like birds of prey.

Then, according to an article in the *Dallas Morning News* by reporter S. E. Hayden (April 19, 1897), two days earlier, on the 17th, one of the strange flying craft crashed onto a property owned by a local judge, J. S. Proctor, after it ran into and was caught in the blades of Judge Proctor's wind-powered water pump atop his well. The craft, reportedly, then exploded into pieces, some of which were scattered across the property and others fell into the well. Actually, as History Channel's *UFO Hunters* found out when they visited the Proctor ranch and inspected the well themselves, the well beneath the wind-driven pump had been filled in almost a century earlier.[1]

After the explosion, Judge Proctor himself reportedly drank from the well and was said to have experienced a terrible reaction to the water. His hands blew up in size until they looked like fleshy bear claws. Was Judge Proctor poisoned by water polluted by an extraterrestrial craft? Was the Texas Airship Mystery really the crash of an extraterrestrial vehicle? The UFO Hunters investigated that and here's what they found.

First, airships, which were lighter-than-air craft and powered by small gasoline combustion engines, were popular sportsmen's adventures in the last quarter of the nineteenth century. In fact, the first president to evaluate the use of a lighter-than-air vehicle for military purposes was Abraham Lincoln, who was pressed to used them as surveillance craft. By the time McKinley took office, lighter-than-air vehicles were, if not commonplace, certainly in use.

Second, as the United States ramped up for the Spanish American War, those wealthy businesspeople and landowners in the American southwest sought to fly some of their vehicles to Cuba to wreak havoc on the Spanish military there.

Third, a quick survey of newspaper articles from the last decade of the nineteenth century in both Earth and Wise counties, Texas, mentioned how folks in the area marveled at the airships. In fact, one historian pointed out to the UFO Hunters that just days before the Aurora crash on the Proctor property, when one of the metal-clad airships had engine trouble, its pilot put it down just outside of Dublin, Texas, near Stephenville, and crowds of locals came out to get a closer look. Was this the craft that crashed onto Judge Proctor's property?

Fourth, what of the metal bits that were spread across the property after the explosion? The UFO Hunters found some of them pressed into the trunk of a tree, and found some of them in the well, which many UFO skeptics said did not exist. The UFO Hunters took the metal pieces to a metallurgy lab at a local university where analysts discovered that the pieces were aluminum, but aluminum that was a pre-1940s alloy. In other words, there was metal where metal should not have been, but it was a traditional substance that was only fabricated before World War II.

Fifth, what about the mysterious ailment, the swollen hands that plagued Judge Proctor after he reportedly drank water from the well, and why was the well closed up? Tim Oates, a descendant of the original ranch owners, explained that Judge Proctor suffered from a form of arthritis and gout that made his hands swell up. It had nothing to do with an alien substance in the well water. And the well itself had been closed up, and the supply pipes pinched off, because the well was going dry and the pump was pulling up silt.

Sixth, what about the body of the alien that was purportedly removed from the wreckage and buried in an unmarked grave? Again, the UFO Hunters visited the graveyard and, although they did not receive permission to exhume any remains from the grave, from which the headstone had been removed years earlier, using ground penetrating radar they were able to ascertain that there were bones buried there, human bones.

So much for the Aurora Texas Airship Mystery. It was a metal-clad lighter-than-air ship.

But President McKinley's ghost may still roam the halls of his former residence and maybe even President Trump will encounter him some dark and stormy night in the White House family quarters.

Chapter 6

TEDDY ROOSEVELT AND THE UFO SIGHTING OVER SAGAMORE HILL

Accrding to Robert Spearing at www.worldufowatch.com, and reprinted by Major George Filer (USAF, Ret.) in his *Filer's Files*,[1] in August 1907, during Roosevelt's second term, folks in the area around Oyster Bay, New York, at around ten at night, a hovering, motionless white orb hung over the president's rooftop. It stayed there for almost an hour. This was not the moon, because it was in the wrong place and too close to the roof, nor was it a star or the planet Venus, because it was too big. According to an article from an old newspaper, *D.C. Evening Star*, cited by George Filer at his Filer's Files website,[2] the light didn't fly away. It simply hung brightly like "an intense white light" over the Roosevelt house and then began to fade at eleven p.m. until it was only a "spark" and then it "extinguished."[3]

At the time, 1907, folks who read about the event or who saw it with their own eyes viewed it as an omen. Why? First, could it have been an omen to President Roosevelt, who was known to spend his summers at Sagamore Hill whenever he could get away from Washington. Therefore, it has been suggested, inasmuch as this light hovered over Sagamore Hill on August 1, a Thursday night, it was as likely as not that if Teddy Roosevelt wanted to get away from the White House for a three-day weekend in the middle of the summer, he might have left the District and arrived on Long Island at night. And if he had arrived home before 11, the light would have still been there. What might it have portended?

According to the Mutual UFO Network's Major George Filer,[4] Roosevelt at the time of the UFO sighting was working on an address to Congress to ask for funding to build a great navy, actually the world's most powerful navy, in part to offset Japan's growing naval power in the Pacific. The Japanese military theory—which other island nations, especially the UK, adhered to—was that the way to defend the country was with a powerful navy that could deter any invasion. The Japanese imperial war machine also relied on its navy to project its military power because Japanese military strategy was premised upon winning wars at sea.

Roosevelt specifically warned Congress about the growing imperial power of Japan in the western Pacific, and of China, and of the growing power of Germany, especially after the Franco-Prussian war. He was prophetic about the need for a modern, powerful American navy to protect the US from threats emanating from the Atlantic and the Pacific. Also, another of his recommendations was the improvement of the facility at Pearl Harbor to become a central port for our navy of the Pacific, specifically a naval base that could host the deep-draft heavy warships required to protect American interests in Asia. Was the floating orb over the Roosevelt home at Sagamore Hill a type of beacon, affirming what Teddy Roosevelt envisioned as a threat to the United States? Or, better, was the orb, which floated there for almost two hours, actually communicating with Teddy Roosevelt, showing him visions of a future in which a powerful United States Navy would defend the nation's interest all over the world even during a brutal war destined to occur thirty-two years into the future and ultimately turn the United States into the world's superpower?[5] Was this event like the orb over Valley Forge, encouraging Washington to keep his faith and to pursue the war to its successful conclusion? Or was this similar to Washington's appearance before a sleeping McClellan, laying out for him the living map of the entire upcoming war between the states?[6]

George Filer also reminds his readers that in that same year of 1907, Theodore Roosevelt's dream of a modern concordat of nations took place at the Hague, according to Filer, "to codify the rules of war—particularly naval warfare." This convention also helped promote the United States from a nineteenth-century isolated nation more concerned with its own internal strife than anything else into a power able to wield influence on the world's stage. Could this, too, have taken place as a result of the communication between whatever intelligence guided that orb and President Theodore Roosevelt? Was it the sighting of a UFO over his homestead that sparked Teddy Roosevelt's vision of an America strong enough to protect the planet, or was President Roosevelt simply looking at the war-making technology of the new twentieth century inspired by Thomas Edison and Nikola Tesla?

Chapter 7

FDR'S PRESIDENCY, WORLD WAR II, AND THE UFO FLAP OVER EUROPE

"A greater power than mine has willed it. I only saw it."

American psychic Jeane Dixon

It would not be far-fetched to assert that the presidency of Franklin Delano Roosevelt marked the beginning of the modern UFO sighting phenomena from the first foo fighter sightings during World War II, the popularly-called Philadelphia Experiment, to the surveillance of UFOs over our nuclear enrichment facilities in the states of Washington and Tennessee. Although by the time FDR was running for his fourth term in the White House, pressured to do so by the Democratic Party and by a war-weary nation not willing to turn to a new commander in chief, his doctors had given him what amounted to nothing less than a death sentence. They told him that his heart was giving out and that he would not last through his fourth term. But FDR, though patrician by birth, was a pragmatic, albeit sensitive, politician who hd navigated America through the Great Depression by avoiding the Scylla and Charybdis shoals of National Socialism on the one hand and Socialism-cum-Communism on the other hand through New Deal legislation that created an entirely new economic paradigm in the United States.

FDR had weathered an assassination attempt and a planned coup d'état, all during his first term. And the American ambassador to the Court of Saint James, Joseph P. Kennedy, partner of mob boss Meyer Lansky during Prohibition and father of future president JFK, had advocated that both Britain and the US side with Adolf Hitler against the Soviet leader Josef Stalin. But FDR had his own vision. When he saw that America would be

threatened in the Pacific by the militancy of the Imperial Japanese shogu-
nate and across the Atlantic by the prospect of Nazi Germany's dominance
of Europe, he skillfully goaded the Japanese into war with the United States
with an oil embargo and a guerrilla war in Burma the kept Japan out of the
Pacific oil fields. He ordered the navy to absorb the Japanese attack on Pearl
Harbor, after having first ordered the aircraft carriers out of harm's way, and
righteously asked a reluctant, but now outraged, Congress for a declaration
of war against Japan. Hitler obligingly declared war on the United States,
and, thus, FDR got his wish and we entered World War II. Ultimately, at the
end of his third tern, though pressed by his doctors to retire, he acquiesced
to party leaders and his own advisors and ran again, and in the final year
of his presidency he bore firsthand witness to extraterrestrial craft and the
beings that navigated them.

The Philadelphia Experiment

It's a common belief that stealth technology is a modern form of advanced
weaponry whose development started late in the twentieth century. But in
actuality, a version of stealth technology, a form of electronic cloaking, was
actively being tested in 1943 and was based upon a methodology developed
during the war by Canada and Britain as a way to defeat German underwater mines. But the true story of what happened in the Philadelphia Navy
Yard in 1943 has been shrouded in mystery and lore even as it made its way
into a series of motion pictures and stories of mind-bending experiments
in time travel and teleportation that involved the Nikola Tesla's old facility near Montauk, Long Island. But the true story of the events which took
place during President Roosevelt's third term in office, and the stories of
the naval officer who investigated them as well as the investigations by cel-
ebrated UFO researchers Ivan T. Sanderson and Morris K. Jessup, was more
mundane.[1]

As America's Pacific naval force closed in on the Japanese home
islands, it was becoming clear that the stubborn enemy was prepared to
take a heavy toll when the time came for invasion. The Japanese high command had mined all of the home waters with devices that would be set off
by the electromagnetic envelopes around warships' hulls, thus keeping
them from any swift entree into the home waters. There had to be a way to

defeat those underwater proximity fuses so that the American fleet's mine-sweepers could clear the mines and allow the rest of the fleet to get inshore for the landing of the ground forces.

In order to cloak their warships' electromagnetic signatures, the navy's plan, according to what Admiral Arleigh Burke told former army intelligence officer and author Philip Corso,[2] the navy at the Philadelphia Navy Yard experimented with giant electronic coils to degauss the hull of the destroyer escort the USS *Eldridge* to reduce the intensity of and shrink its hull's electromagnetic envelope. However, when the vessel was run through the coils, the heat generated by the amount of electricity was so intense that not only did it begin to melt the vessel's superstructure, two sailors among the crew were actually fused to the deck. Because this experiment was key to the development of the navy's stealth technology, an electronic invisibility cloak, it was top secret, and when the sailors were killed, the vessel was dispatched down the Intracoastal Waterway to the naval shipyard at Newport News, Virginia, as a fog bank closed in. It was such an eerie scene as clouds shrouded the *Eldridge* that Carl Allen, a crew member on another vessel, wrote to author Morris K. Jessup to reveal what he described as the secret of the ship's teleportation and time travel. But the whole story, although it was a hoax, captured the interest of Ivan T. Sanderson and, years later, prompted the navy to investigate the entire incident through the Office of Naval Research, according to Jan Hester writing in the late great *UFO Magazine*.[3] The ONR tasked officer George Hoover to investigate the stories surrounding the USS *Eldridge*'s disappearance from the Philadelphia Navy Yard, but Hoover determined that time travel was not the cause of the vessel's disappearance. However, Hoover was so intrigued by the possibilities of time travel that undertook his own research into the possibilities before he died, discovering that the US military was well aware of the presence not only of UFOs, but that the ETs who navigated them were us from the future.[4]

The tasks of winning final victories in Europe and in the Pacific and of dropping the top secret atomic bomb would be left to his vice president Harry Truman, who, as we shall see, reported his own hearing of the ghostly footsteps of presidents James Buchanan and Franklin Pierce and would be the first president to come face-to-face with the wreckage of a UFO.

FDR and UFOs: The Alien Bodies

According to Grant Cameron's excellent research on the Presidents UFO website,[5] the Roosevelt Administration already had preserved bodies of extraterrestrials in its possession at the Pentagon as early as 1939. Based on information from CUFOS, the Center for UFO Studies, Cameron cited a letter from a descendant of Senator, then Secretary of State and 1945 Nobel Peace Prize winner Cordell Hull, which said, in part, that Hull swore his cousin, whose daughter submitted the story, which wound up in CUFOS, to secrecy because it was classified. The letter said that Cordell Hull showed his cousin an "amazing sight." He took him, the letter said, "to a sub-basement in the U.S. Capitol building, and showed him an amazing sight:

1. Four large glass jars holding 4 creatures unknown to my father or Cordell [and],
2. A wrecked round craft of some kind nearby."[6]

The interesting aspect of this revelation is that it, in part, conforms to what Philip Corso said about his experience when he was brought to the Pentagon by Lieutenant General Arthur Trudeau early in the Kennedy administration as the head of the Foreign Technologies Division of Army R&D, a story we will cover later. But if the Cordell Hull story is true and the Roosevelt administration did have alien artifacts in its possession as early as 1939, then what happened on February 25, 1942, right after the start of America's entry into World War II is even more stunning. And that story concerns the well-reported—and heavily covered in the news—Battle of Los Angeles, footage of which can be seen on YouTube.[7]

The Battle of Los Angeles

There are still those who argue that the strange aircraft flying over California South Bay, out over the Pacific, was nothing more than a weather balloon that had broken loose from its tether just after midnight on February 25, 1942, and spooked all of the antiaircraft gunners along the shore who picked it up with their searchlights and fired round after round of shrapnel-laden flak at the object. But they couldn't bring it down. In fact the arguments about it continue to this day even though the aspects of that event defy conventional explanation.[8]

After the attack on Pearl Harbor on December 7 the previous year, cities along America's Pacific coast were on alert because there was a fear that the Japanese would launch attacks against them. That fear became palpable when a fleet of Japanese submarines sank a number of American vessels off the West Coast from Alaska to Mexico. The Japanese landed a small force of infantry on the Aleutian Islands early in the war, killing some American civilians and taking prisoners. A Japanese submarine also attacked oil refinery facilities near Santa Barbara and, although leaving part of the facility in flames, no other damage was reported and there were no American casualties. Needless to say, reports of Japanese naval activity off the West Coast was unnerving both to the military and to residents all along the coast from Washington State to San Diego, California. Therefore, the sighting of an object flying over the Pacific just off the coast of California was enough to set off alarms among the antiaircraft batteries set up to protect the US from enemy bombers.

The object the gunners sighted and which was pinpointed by shore-based searchlights seemed to float slowly southward along Santa Monica Bay to Long Beach, and although shells were exploding all around it, lighting up the night sky, the object seemed not to have been hit. Either those aiming the antiaircraft guns were less than competent or the object itself seemed to be able to withstand near hits or the object was flying at too high an altitude for the shells to reach. As shells missed their targets and exploded in midair, some of them fell back to earth along the beaches and pieces of hot shrapnel started fires in some structures near the shore. Other bits of shrapnel lodged themselves in the soft sand and were picked up by residents as well as LAPD officers sent to the area on the following morning.

The flying object withstood the tremendous bombardment and seemed to float away, disappearing from view, which is when the shelling stopped. Both military and civilian authorities launched investigations, after which the army said that it was clear to them that the object was a balloon that was mistaken for an aircraft. But no one could explain why the object was not hit even though it was moving slowly and had been captured by the searchlights and was in the gunners' sights as it moved along the shore. In a secret letter from General George Marshall in the War Department to President Roosevelt, Marshall explained that the mysterious object, or objects, were actually airplanes, probably non-military, that, in Marshall's words, "were operated by enemy agents for purposes of spreading alarm, disclosing locations of antiaircraft positions, and slowing production through blackout." Marshall assures the president that the aircraft traveled at "varying speeds" and that "no bombs were dropped."[9]

For his part, President Roosevelt seemed assured that these objects were conventional aircraft, but, he wrote back, "who is responsible for the air alarm system in the United States?"[10] He seemed perturbed that civilian authorities had decided to fire on the objects and asked that all air alarm response be placed under the province of the War Department. The Battle of LA is still a mystery, even though Marshall's letter seems to resolve it. But, although it is still one of many regarding unidentified flying objects in FDR's administration that remain unsolved, at least a few of the eyewitnesses to the events said they could clearly see a flight of airplanes caught in the searchlight beams. Whose were they?

We believe that the planes were ours. Why? Because that night, after the events in Alaska and along California's coast, when America's war fears of an invasion were at their height and still months away from the Doolittle raid on Tokyo, the American military decided to "mystery shop" our western coastal defenses, the response of our antiaircraft artillery crews, and even the range of the shells they were firing, by sending a small group of conventional aircraft along the coast at an altitude beyond the range of our antiaircraft artillery. Just the way we did during the U-2 flyovers of the USSR, American fliers were sent up along the California coast that night to see what the response would be and whether the mobilization of coastal antiaircraft could defend against Japanese bombers. That was the reason there were no US Army Air Force interceptors in the sky that night, allowing the testing aircraft to move freely down the coast.

There might have been another, and far more nefarious, reason for the entire Battle of LA incident, which involved whipping up war fears among the American public so that the internment of Japanese American citizens would not inspire a general protest from the public. It may sound incredible that even while the world was witnessing the Nazi concentration camp horror, the American public acquiesced to what was nothing less than an unconstitutional mass imprisonment of Japanese American citizens and even their children. If it is true that the Battle of LA was a type of false flag test of our shore defenses for which part of the rationale was to make it publicly acceptable to incarcerate Japanese American citizens en masse, and their children, too, then it is not very different from the rationale behind the Reichstag fire, started by the Nazis, and the resulting Krystallnacht, night of the broken glass, when Jews were attacked in the streets of Berlin and other German cities.

The "War of the Worlds" Radio Broadcast

Even before the Battle of Los Angeles, FDR's War Department knew how easy it would be to control large populations through the fear of attack. FDR had witnessed this firsthand four years earlier when, again with the help of the Rockefeller Foundation, the Mercury Theatre on the Air broadcast a reality radio show in the form of a news broadcast based upon the H. G. Wells novel "War of the Worlds" on October 30, 1938, which you can listen to online.[11] And, no, Orson Welles and H. G. Wells were likely not related.

The radio show was broadcast over the Columbia Broadcasting System, CBS, but was put together by Frank Stanton, an expert in the very early science of statistical analytics to measure public perception, who was working at Princeton University at the time and who would eventually become the head of CBS. It's no accident that the landing of the Martians took place in the town of Grovers Mills just outside of Princeton near Cranbury, New Jersey. Frank Stanton used the Orson Welles broadcast as part of a test to see whether an event and follow-up publicity about panic in the streets could stir up people into a frenzy. The "War of the Worlds" broadcast was later used by the RAND study to substantiate the government's claim, a claim still in force today, that the revelation of extraterrestrials coming to earth would set off panic in the streets. But, first, there was no real panic in the streets because, as Frank Stanton himself said, "In the first place, most people didn't hear the show."[12] Talk about "fake news"! These guys invented it. Second, there was a far more nefarious aspect to the "War of the Worlds" broadcast, an experiment to see whether mass populations could be controlled to manipulate opinion to the point where the manipulators could get away with anything. It was the run-up to FDR's decision to sit for the attack on Pearl Harbor to get the US into World War II, which was followed by the Battle of Los Angeles to whip the public into a greater frenzy against the Japanese Imperial forces so that the public would not resist the incarceration of Japanese American citizens. This was FDR's version of the Reichstag fire.

Foo Fighters Over Europe and the South Pacific

As the United States Army Air Force pressed the war against Germany in 1944, fighter pilots became aware that they were being followed, even through some of their maneuvers, by balls of red light, seeming on fire, that moved with them. Because the balls looked like hanging balls of fire, the pilots used the French term, "feu," for fighting fireballs. The name was pronounced "foo," and the name "foo fighters" became the slang terms for the objects. No one knew what they were even though some scientists believed that they were balls of lightning, like the St. Elmo's fire that alighted near Captain Ahab in Melville's *Moby Dick*.

But these weren't balls of lightning because they seemed to organize themselves in formations and maneuvered with, not independent of, the American fighters escorting heavy bombers over Europe. The next theory that the Allied war departments settled on was that these balls of light were German secret weapons. And that would have made sense because in the latter years of the war, the Germans began introducing some of their experimental weapons into the fight. American fighters had already encountered the jet-engine Messerschmitts, the Me262, which could outmaneuver the conventional Allied propeller-driven fighters. But these foo fighters were different. They were clearly not conventional aircraft, they had no wings and no obvious source of propulsion, yet at the same time they seemed to be navigated with a degree of intelligence. They flew with the Allied fighter formations without interfering with them, almost as if they were surveillance devices keeping tabs on the progress of the air war.

Theories about secret German weapons were rampant during the latter years of World War II, and were bolstered by actual sightings of Nazi rocket planes that dived down upon Allied fighter and bomber formations from very high altitudes. But the Germans did not produce enough of them to have any real impact on the Allied bombing raids, nor were the rocket planes able to engage American fighters effectively enough to deter the formations from protecting the bombers on their missions. The presence of these rocket planes, however, did lead to Allied theories that the foo fighters were secret Nazi remote-controlled weapons even as Luftwaffe officers came to believe that the foo fighters were secret British or American weapons. Neither side was correct, and foo fighters—some people today call them orange orbs—remain a mystery of the World War II and FDR years.

The presence of foo fighters did not go unnoticed by the media, which reported these anomalies as early as 1941 and as late as 1960. In at least two instances, pilots of a B-25 bomber and a B-29 strategic bomber sighted these balls of light not only over Europe, but over the South Pacific as well. In one instance, a B-29 gunner opened fire on a foo fighter over the Pacific and the object split into a cloud of flaming pieces.

UFOs Over Nuclear Enrichment Facilities

The development of American nuclear weaponry and the race between the Allies and the Axis to deploy nuclear weapons is a story that could fill more than one volume of history. Suffice it to say that from early in the war, American nuclear scientists believed that if either the Germans or the Japanese got to the nuclear finish line first, the war would be effectively over. For that reason, with the deepest secrecy and under the code name the Manhattan Project, American scientists sought to be the first to miniaturize and enclose a nuclear stockpile and develop a fuse mechanism that would detonate a nuclear reaction, all of which could be packaged into a single bomb capable of being dropped by a heavy bomber. Of course, the Germans were actively at work on nuclear weapons, too, and the Japanese had constructed a nuclear enrichment facility at Konan in what is now North Korea, and in 1945 had actually tested a nuclear detonation, all of this according to journalist Robert Neff in the *Korea Times*.[13] It should be noted that the United States Air Force destroyed this facility at the outset of the Korean War, flattening it completely in one of its early bombing runs.

The Nazis, too, were working on the formula to detonate a critical mass to create a nuclear explosion. That, combined with the development of the V-1 and V-2 and the threat that U-boats with missile launchers on their foredecks would steal themselves into America's coastal waters, turned out to be one of the strategies the German high command was trying to perfect after the collapse of Operation Barbarossa on the Eastern Front. Because the Luftwaffe did not have an intercontinental bomber, their plan was to launch guided and cruise missiles from U-boats, one of which craft still sits today at the bottom of Long Island Sound off Port Jefferson, its insignia and weapons having long been removed by the US Coast Guard, a silent rusting testament to what might have been. If the Germans could inflict enough damage on the United States to force it out of the war or to seek a separate peace, that would have allowed Nazis to focus all their efforts on keeping the Allies out of Germany and allowing the Reich to remain in power. But it was not to be.

As documented by Jeremy Bernstein in *Hitler's Uranium Club: The Secret Recordings at Farm Hall*,[14] the Germans were all but advanced to the point of developing the fuse trigger for a nuclear bomb, the conventional explosion inside the bomb to push together the elements of nuclear fuel to start an uncontrolled chain reaction resulting in an explosion. The recordings at Farm Hall, Cambridge, reveal that as the German scientists at the war's end wondered why the Nazis never developed the trigger, others in the room acknowledged that though the Nazi nuclear weapons research was distributed among different institutes, Werner Heisenberg explained that he had solved the trigger formula, but kept it from Hitler and the Nazi high command.

Further evidence that the Germans were developing a nuclear weapon also comes from the story of an American spy in the OSS, Professor Morris Berg, a Romance language lecturer at Princeton University and an all-star major league baseball catcher. This was one of the great stories coming out of FDR's War Department and the American spy apparatus keeping tabs on Nazi weapons development during the war. As documented in Nicholas Dawidoff's *The Catcher was a Spy: The Mysterious Life of Moe Berg*,[15] soon to be a motion picture, Berg wasn't just a baseball player and a college professor. He was dispatched to Europe by the Manhattan Project's General Leslie Groves to attend a conference in Switzerland to hear Heisenberg speak about nuclear weapons and to determine whether Heisenberg had worked up the formula for a bomb. If, in his opinion, Heisenberg held the key, then it was Berg's job to kill him and bite down on the poison pill tucked into the side of his mouth lest he be captured and sent to a concentration camp. Berg was Jewish. As it turned out, Berg determined that Heisenberg did not have the formula, did not kill him, and was secreted out of Europe to the United States.

Here in the United States, reports of UFO activity in 1945 over both the Hanford nuclear enrichment facility in Washington and The Oak Ridge nuclear enrichment facility in Tennessee indicated that, if the reports were true and UFOs were surveilling our advanced nuclear weapons research facilities, then these events might have been the set up for the crash at Roswell during the Truman administration two years later. These reports suggest that whoever, or whatever, was navigating these unexplained unidentified craft, a formation of which were spotted by Kenneth Arnold over Mount Rainier in 1947, that surveilling entity was very interested in the human development of an energy source which, if unleashed in a multi-lateral nuclear exchange, had the potential of wiping out most life on planet Earth, especially human life.

Die Glocke die Wunderwaffe war

Given the nature of FDR's wartime intelligence apparatus, it is likely that he would have been briefed about a top secret Nazi project tucked away inside a hollowed-out cave in Poland's Owl Mountains. This was a secret project those who knew about it in the Nazi weapons community called "the wonder weapon," the one device that could end the war even as Roosevelt's, Churchill's, and Stalin's armies closed in on the Third Reich. This was a giant bell-shaped device being assembled, and presumably tested, underground where Allied aerial surveillance could not find it.[16] It was inside this mountain that Nazi scientists, utilizing prisoners from a nearby concentration camp, prisoners whose lives were sacrificed for the top secret weapon the Germans were developing, were experimenting with a high-radiation bell-shaped device that, reportedly, was so powerful that it had the ability to change the outcome of the war even in its final months.

This object was referred to by the SS officers overseeing its development as a "wonder weapon." They called it "the Bell" or, in German, "Die Glocke." There is a lot of speculation about this device. But the story that many have come to believe is that the weapon was composed of two counterrotating cylinders containing a liquid heavy-metal substance referred to in the Hindu Vedic texts as red mercury, a substance that when spun at high speeds was highly radioactive. The counterrotation generated a radioactive energy field, the story goes, that could propel the Bell great distances not only through space, but through time. Now imagine a device that could transport itself back to Washington, DC, circa 1940, where it would detonate, wiping out the United States government and rendering the US helpless in joining the war. Imagine such a device detonating in London in early 1939 or in Moscow in 1940. Hitler would have been in control of all of Europe.

The radioactive energy field from the experiments with this device was so powerful that hundreds of workers in the cave were irradiated and killed, even some of the Nazi scientists. Now imagine that this project, managed by an SS colonel named Kurt Debus—an electrical engineer—simply disappeared from inside the cave as the Soviet army smashed through Poland, overwhelming the German Wehrmacht. Where did it go? Now imagine that that very same SS officer Kurt Debus turns up at Cape Kennedy in 1965, during the Lyndon B. Johnson administration, as a launch director and NASA director. Now imagine even further that a bell-shaped object glowing a purplish blue crash-lands in a ditch in Kecksburg, western Pennsylvania, and is retrieved by an army unit waiting for it, protected not only by the military but possibly by security personnel from NASA and even from a US intelligence agency. Why was NASA at this crash site? A question for the late NASA official Kurt Debus waiting for his precious Wonder Weapon to return to its developer. Could that device, after landing in Pennsylvania in 1965, have made its way further to the waiting arms of a unit from Area 51 after it crash-landed in Needles in 2008 during the last months of the George W. Bush administration? And why would the Nazis, in their attempts to develop such sophisticated technologies as ballistic missiles, rocket planes, and jet engines, have focused so much of their attention upon the Vedic texts to find, among other things, the concept of Die Glocke? The answer, you may be surprised to learn, lies in a phenomenon called the Great Age of Spiritualism that spread across Europe and the United States in the late nineteenth century, which informed part of the Nazi philosophy of the origins of Aryanism.

The seemingly outrageous theory of a World War II weapon that could travel through time, as far-fetched as it may seem, is what still drives UFO historians to determine the nature of the Nazi secret weapons program and how the very scientists who developed some of those sophisticated weapons were patriated to the United States at the very beginning of the Truman administration to compete with the Soviets to develop aerospace technology.

The Nazi Bell, rather than a time machine, as exciting as that theory is, was more than likely a nuclear enrichment device to purify fuel for what the Germans hoped would be the super weapon that could force the Allies to sue for peace. The Roosevelt administration had been warned by expatriate European scientists working in the United States that the Germans were seriously experimenting with nuclear fission. The Bell was likely based in part on descriptions from the Vedic texts of the god-like super beings and the craft they flew called the "Vimana" that fired death rays at their enemies and, some say, were early descriptions of nuclear weapons. Is this what the Germans were developing? And if so, where did the device go when the Soviet troops reached the Owl Mountains? We know that Kurt Debus came to the United States under Operation Paperclip at the end of the war and became, along with over a thousand other German aerospace engineers, one of the leaders of America's space program.

Eleanor Roosevelt's Belief in UFOs

There were enough UFO stories circulating around the White House and the Department of War during the Roosevelt administration to pique the then First Lady's interest in the entire subject. But, as First Lady, and with enough on FDR's plate in terms of fighting the war in Europe and in the South Pacific, the last thing the country needed was a press report that she had a belief in presence of extraterrestrials. However, after the news broke—and then was retracted—in July 1947 that the Army Air Force had gotten its hands on a crashed flying saucer outside of Roswell, New Mexico, and with the formation of a top secret study group to study the UFO phenomena and make reports to the president about the nature of these phenomena, the former First Lady began to take a stronger interest in the subject.

Also, because early in the twentieth century, a circle of Russian women, self-professed psychics, believed that there was a mystical force called "the Vril" that was an empowering human force that flowed through the universe and that human beings who could utilize this force psychically were thus empowered, early Nazi propagandists sought the origins of this force. Combining the theory of the Vril with Helena Blavatsky's belief that the human race, that is, those who were Caucasian, came to Earth from the planet Aldebaran and the Aryans from that planet initially settled in Thule, the Vril Society ultimately turned into the Thule Society and became the quasi-philosophical basis for myth of the Aryan übermensch. And we know how that turned out.

Eleanor Roosevelt, by the 1950s, was fascinated by these theories of an alien race that came to planet Earth to become the basis of humanity and, through the exploration of linguistic texts, encouraged scholars to relate the nature of the linguistic texts to the ancient belief in the presence of extraterrestrials on earth.

We may think that the prejudices, fears, and hateful racially destructive acts of the first half of the twentieth century, and the distorted scientific theories behind them, died with the end of World War II, the passage of the Civil Rights and the Voting Rights acts under President Johnson, and maybe we still think that today. However, in light of the 2016 presidential election, think again.

Chapter 8

HARRY TRUMAN: ROSWELL, THE SUMMER OF THE SAUCERS, AND MK-ULTRA

The UFO Crash at Roswell

It would not be too far-fetched to say that one of the most stunning events during the Truman administration was the Roswell incident that changed American history and quite possibly our understanding of human evolution. Whether it was a crash landing of an actual flying saucer, or flying crescent similar to the one Kenneth Arnold said he spotted over Mount Rainier, or a craft of another shape, or the crash of something else, documents from the period seem to indicate that United States government agencies were more than curious about it. You can see some of the relevant official correspondences among the agencies from the declassified and released FBI files online.[1] UFO historian and former navy photo analyst Dr. Bruce Maccabee has also included these files in his new book, *The FBI-CIA-UFO Connection: The Hidden UFO Activities of USA Intelligence Agencies.*[2]

We know that something did happen at Roswell because of all the fuss. There was that press release saying the army had captured a flying disk outside of town. Then there was that story the next day featuring a smiling Major Jesse Marcel in front of a weather balloon under a headline saying that the army had retracted its statement because it was really just a standard off-the-shelf weather balloon. Of course, that made no sense because, as Stanton Friedman has pointed out many times, especially on *UFO Hunters,* that any Army Air Force officers at the 509th bomber wing at the Roswell Army Air Force base would have recognized the crash debris from a weather balloon immediately because not only were the weather balloons assembled right across the street from the base, but they were fabricated out of neoprene rubber, the very same material used throughout World War II. The weather balloon story was fake news, a cover-up.

But then there was Jesse Marcel himself, popping up almost thirty years later, saying something did happen at Roswell. And then, years after that, the air force came out with another statement saying that it really wasn't a weather balloon after all. It was a super secret Project Mogul balloon, made of exotic material, and that's why it fooled people at the time. Project Mogul was a high-altitude device designed to sniff the atmosphere for evidence of the Soviet Union's atomic bomb tests. But that didn't satisfy the curious people researching the Roswell story because there were all those witnesses, those pesky witnesses who said they remembered exactly what they were doing in Roswell during those weeks in July 1947 and what the Army Air Force told them. Many of these were eyewitnesses who even said they saw strange-looking little creatures being transported to the Army Air Force base at Walker Field on the edge of the city of Roswell.

Then, fifty years to the month after the reported Roswell incident, the air force released yet another report saying that the entire incident was really a case of crash dummies that crashed in the desert outside of Roswell. Case closed. Just a bunch of dummies that other dummies thought were alien creatures. Too bad the crash dummies were so burned up in the crash. They were great-looking dummies, photos of which would have resolved the whole problem. The crash dummy explanation might have had some traction had there not been a tiny problem: not only were the crash dummies too small and not properly sized adults, but crash dummies weren't used until the early 1950s. The Roswell incident took place in 1947. The air force would have had to have sent the dummies back in a time machine.

Moreover, the crash dummy story didn't really satisfy anyone researching the case because of all those reported government documents doing what documents do, documenting the entire affair. The documents, called the "MJ-12" documents,[3] purportedly revealed the truth about the crash, saying it was actually an alien spacecraft that contained real ETs—extraterrestrial biological entities, or EBEs, they called it—and the whole event really happened the way the researchers said it happened. The documents even contained briefing information for the incoming president, Dwight Eisenhower. Then, as expected, there was another dispute over the documents.

"They were planted," debunkers said, "simply to throw researchers off the trail. They were hoaxes." But researchers were undeterred and investigated the documents and said they comported with statements made by Army Air Force and government officials at the time, comported with the style and presentation of documents at the time, and were real, not hoaxes.

Next came the claims of retired Army Lieutenant Colonel Philip Corso, who said that he actually saw the Roswell alien at Fort Riley in 1947.[4] Then the debunkers jumped on him. Then in 2007, Jesse Marcel Jr. wrote his book, *The Roswell Legacy: The Untold Story of the First Military Officer at the 1947 Crash Site,*[5] about the experiences of the Marcel family and the Roswell story in which he talked about the crash itself, his father's experience retrieving debris from the crash site, and Jesse Marcel Jr.'s own recollections handling the strange crash debris when his father stopped off at the Marcel house on his way back to the base. Jesse Jr. remembers well, as an eleven-year-old, that his father said to him and to his mother to look carefully at the material he had brought back from the crash site because it was something that the family, or most people on earth, for that matter, had never seen before and would likely never see it again. Jesse Marcel Jr.'s retelling of the Roswell incident, particularly in light of his family's experiences, was the coup de grace to the military's Project Mogul theory in which they argued that what really fell out of the sky was a top secret balloon device to sniff out the remnants of Soviet nuclear testing in the atmosphere.

Perhaps one of the most revealing books about Roswell was Thomas Carey's and Donald Schmitt's *Witness to Roswell* in which, as the book title indicates, the authors go back to the very beginning of the case and interview all of the witnesses, including Lieutenant Walter Haut (Ret.), the public information officer at the Roswell Army Airfield in 1947, about what he was told to do by his commanding officer Colonel William H. Blanchard. Haut was the officer who delivered the news to local newspapers that the army had captured a flying disc outside of Roswell, a story that made the headlines all over the country before it was retracted the next day by the army.[6] In *Witness to Roswell*, Schmitt and Carey reprint a statement, functionally an affidavit Haut swore out before an attorney and a notary in contemplation of his death—authorizing his family to release it at their own discretion after his death—in which Haut affirms that he was physically present at the debris field, saw the wreckage of the UFO with his own eyes, saw the bodies of aliens on stretchers, saw the craft under a tarp on a flatbed truck, held a piece of the strange debris from the crashed object in his own hands, tried to keep a piece of the craft on his desk as a kind of souvenir, and was part of a conference of officers at the Roswell Army Airfield in which the mechanism of the cover-up was discussed and put into place. The late Walter Haut's statement in contemplation of his death can be found at http:// roswellproof.homestead.com/haut.html#anchor_8. Although Haut's statement has been criticized as only being responsive to questions posed by Schmitt and Carey, there is extrinsic evidence substantiating Haut's revelations, statements he made in a videotaped interview prior to his affidavit that he was indeed at the crash site. Haut was also reported to have told visitors to the UFO Museum in Roswell that he was at the crash site and saw the crashed UFO, which revelation prompted local UFO historians Dennis Balthaser and Wendy Connor to memorialize his statements on videotape. The extrinsic evidence corroborates Walter Haut's statement to his attorney about his role in the cover-up of the UFO crash at Roswell. Since the publication of *Witness to Roswell*, Carey and Schmitt, as well as one of the original Roswell incident researchers, Kevin Randle, PhD, have decided to review all of the information about the case again, including what might have happened to pieces of the craft.

Six years ago, Annie Jacobsen, a *Los Angeles Times* editor, in her book about Area 51,[7] taken seriously by news commentators ferociously anxious to debunk anything UFO even though she actually misstated the date of the base's inception, which wasn't 1951, as a simple fact-check would have shown, repeated the same Roswell debunking, also factually inaccurate in her book. She explained that a Horten Brothers flying wing was responsible for all the confusion about what had actually landed. She said that the army had tracked down the Horten brothers after 1947, who admitted they were working with the Russians. The theory was that the Horten brothers craft, which in the years prior to World War II was actually a flying wing, a type of craft depicted in the H. G. Wells 1936 motion picture *Things to Come*, had been dispatched to Roswell by Josef Stalin to frighten America. The craft—and the original Horten flying wings were made of wood, not some exotic compound—even had Cyrillic markings on it.

Imagine a base intelligence officer, who'd been to radar school shortly before the crash, not recognizing Cyrillic or confusing wood with another material, or not knowing what a jet engine was in 1947. Imagine further that while Josef Menegle was on the run through South America to avoid capture by both American and Soviet Nazi hunters while practicing the science of eugenics to produce Aryan babies, he stopped to provide Stalin with deformed children who would burn up in the crash and be confused with extraterrestrials. What's even worse, this entire explanation of the crash of a secret airplane at Roswell is actually a cover story based on the James Blish short story "Tomb Tapper" that appeared in the July 1956 issue of *Astounding Science Fiction*.[8] If all of this sounds implausible, now imagine that all of the pundit heads who embraced this theory because it sounded rational and provided the standard "anything but aliens" answer to stories of UFOs were reporting it without inviting anyone with solid evidence of what actually happened to give an alternate opinion. Almost sounds like a cover-up.

If you want a real cover-up of what might have been a concern when the army detected the crash debris, consider this. During the latter years of World War II, from 1944 to 1945, the Japanese launched what they called "fire balloons" or "fusen bakuden," hydrogen-filled balloons carrying an incendiary device laced with shrapnel over the Pacific to northwestern US cities. Some of these actually landed and caused damage. However, for war censorship purposes and so as not to give the Japanese military information about their success or any propaganda advantage, the US military did not allow reports of these balloon landings to appear in the press. It was a successful censorship operation that is still taught in CIA training today. Thus, upon coming upon debris in New Mexico, perhaps a first thought was that it was a Japanese balloon bomb that had drifted over the Pacific years after the war ended or maybe even a doomsday weapon launched by remnants of the Japanese military. Best to cover it up until they knew what it was. But the Jesse Marcel and Walter Haut stories themselves convincingly contradict the balloon bomb theory as well as the Annie Jacobsen story in her book about Area 51.

In brief, the Roswell story is about the discovery of crash debris forty miles outside of the city of Roswell by a rancher named Mac Brazel, whose story can be found in the Schmitt/Carey *Witness to Roswell* and elsewhere online[9], in which he describes how he found the strange, exotic material spread over the landscape at the Porter Ranch and brought a box of it to

Chaves County Sheriff George Wilcox, who put it in a jail cell for safekeeping and then called the 509th Army Air Force unit at Walker Field at the edge of Roswell. The entire story of the events that transpired when Brazel brought the box of debris from the field at the Porter Ranch to Sheriff Wilcox are documented by his wife, Ines Wilcox, in her unpublished diary about her year tending to inmates at the Chaves County jail. In that diary she wrote:

"One day a rancher north of town brought in what he called a 'FLYING SAUCER.' There had been many reports all over the United States by people who claimed they had seen a FLYING SAUCER. The rumors had many variations. The saucer was from a different planet, and the people flying on it were looking us over. The Germans had invented this strange contraption, a formidable weapon. Other tales, that one had landed and strange looking people all seven feet tall or more walked from it, but quickly departed on sighting any onlooker.

"All the papers played the stories up, and many people searched the skies at night to catch sight of one. Since no one had seen a flying saucer, Mr. Wilcox called headquarters at Walker Air Force Base and reported the find. Before he hung up the telephone, almost, an officer walked in. He quickly loaded the object into a truck and that was the last glimpse alone had of it. Simultaneously the telephone began to ring, long distance calls from newspapers in New York, England, France, government officials, military officials, and the calls kept up for 24 hours straight. They would speak to no one but the Sheriff. However, the officer who picked up the suspicious looking saucer admonished Mr. Wilcox to tell as little as possible about it and refer all calls to Walker Air Force Base. A secret well kept, for to this day, we never found out if this was really a FLYING SAUCER."[10]

A day or so after Brazel's discovery of the debris at the Porter Ranch, a team from the base led by intelligence officer Major Jesse Marcel and Counter Intelligence Corps commanding officer Captain Sheridan Cavitt visited the area and then ordered a team from the base to retrieve the material scattered over the ground, especially what some of the army personnel called "memory metal," a foil-like substance that could be wadded up, crushed, and then when released would pop back into its original shape. The material was brought back to the Army Air Force base outside of Roswell, whereupon commanding officer Colonel William H. Blanchard notified his commanding officer, General Roger Ramey at the Eighth Air Force at Fort Worth, Texas. For the full story of what happened next, Lieutenant Walter Haut, the public information officer for the 509th, explained that the entire cover-up was put in place by General Ramey most likely upon direct orders from the Pentagon.[11] You can also find the complete statement in the Schmitt and Carey *Witness to Roswell*, in which Haut details all of the events that took place after the crash debris was discovered and talks about his own visit to the crash site and his seeing not only the craft, but the little alien bodies on stretchers. Further, Haut reveals in his affidavit made in contemplation of death that after Major Marcel was dispatched to Fort Worth, he was ordered to pose for a photo in front of a weather balloon instead of the Roswell crash debris. When he returned to Roswell, he was a bitter man, who vowed never to speak of the incident again. He said the same thing to his wife and his son Jesse Jr. But in 1978, he went public with the entire story.

The Roswell story didn't just end with the material being shipped to Fort Worth and thence to Wright Field in Dayton, Ohio. One sequel to the story began to slip out on the night of the first *Apollo* moon landing, when one of the children of former Marine Lieutenant Colonel M. "Black Mac" Magruder commented that he wondered whether there were entities, space people, on other planets. Magruder gently revealed to his son that "yes, there are, and I have seen one of them, but that's all I can say." As cryptic as that statement was, it wasn't the last time Marion Magruder would reveal the full story of what happened to him when, as a member of the 1948 class at the National Air War College, he and his class visited Wright Field to see the debris from the crash at Roswell and to view the living ET that had been captured from the crash.

Magruder described the ET to his children as "Squiggly," the name he gave it because of its long arms and the way its body seemed to wave while it sat there in its enclosure. Magruder described it as scary because it looked almost human, with its pinkish hue, but its head was oversized like a large light bulb and its eyes were very large and a deep black. It communicated with him, he said, telling him that his captors were running tests on it that were killing it. Magruder finally blurted out to his children, "this was one of God's creatures and what they did to it was a shameful thing."[12]

The Day After Roswell

What happened to the debris from the UFO that had crashed at Roswell? According to retired Colonel Philip J. Corso, who was second in command to Lieutenant General Arthur Trudeau at Army R&D from 1961 to 1963, some of the army's stash of UFO debris wound up at the Pentagon, where General Trudeau and Senator Strom Thurmond had ordered that a team of scientists there examine the material and determine what military defense contractors were working on similar technology and deliver the UFO debris, along with a working development budget, to those contractors to reverse engineer this material. Out of this cache of material, Corso wrote in *The Day After Roswell*, came advances in laser technology, Kevlar, fiber optic cabling, and the integrated computer circuit.[13]

Corso explained that even as early as 1947, President Truman had asked the army to determine the nature of the lowest common denominator of the technology and deliver it to an American company that was working in that area. The army delivered pieces of what looked like electronic circuitry to Bell Labs, where Brittain and Shockley were working on a conductive switch to pass one electron at a time though the circuit. Unlike the Edison light bulb, just like a radio tube that conducted electricity through a vacuum, the scientists at Bell Labs sought to invent a switch that would be more stable and precise. They had experimented with a circuit in a silicon base, but they could not control the flow of electrons one by one. When, according to Corso, they received the pieces of circuitry from the UFO, they found that by doping their silicon base with arsenic, they could move only one electron at a time across the circuit. This became the basis for what Bell Labs patented in 1948 and afterwards, and it was called the transistor. The development of the transistor, a switch that both transmitted electrons and resisted them, became the foundation for all of the integrated circuitry in use today, especially the computer.[14]

UFOs over Washington and the Summer of the Saucers, 1952

On two successive weekends in July 1952, an echelon formation of circular UFOs overflew Washington, DC, a story heavily covered in the *Washington Post*[15] as well as on Movietone newsreels at the time. But the story is much bigger than simply a sighting. Not only were the craft sighted visually, they also turned up on multiple radar screens, both civilian and military, and were chased out of the area by Air Force pilots. Worse, as author and UFO researcher Frank Feschino has written in *Shoot Them Down* and has explained at various UFO conferences,[16] there was at least one UFO shot down by the air force, which crash-landed in Braxton County, West Virginia, and dispatched a defensive robot—shades of *The Day the Earth Stood Still*—which local residents saw and characterized as the Braxton County Monster. There were multiple witnesses to this, whose statements are documented in Feschino's excellently researched history, including military officers, who were dispatched to secure the area and prevent the story from getting out. News of this crash landing and the events that followed certainly went to President Truman, who was at the very end of his term in office.

In 1956, even Captain Edward J. Ruppelt, the head of Project Blue Book, based at Wright-Patterson Air Force Base, revealed the truth about the 1952 UFO invasion of the skies over Washington in his book *The Report on Unidentified Flying Objects*,[17] revealing that he was warned about the impending flyover by scientists working for the CIA when UFOs turned up over the East Coast of the United States and Chesapeake Bay earlier in the summer. Ruppelt wrote that USAF pilots actually chased flying saucers, tried to lock their radar-directed guns on them, and even engaged in dogfights with them, but they were outmaneuvered at every turn. When the time came for the air force to explain to the American public what the nature of these mysterious objects were, then General Samford said that the whole thing was a misidentification and a misunderstanding. There were no UFOs, only odd radar images and false observations, anomalous illusions, by the pilots.[18]

According to the late Philip Corso and to briefing documents for the incoming President Eisenhower in 1953, after FDR, Harry Truman was the next president to be presented with hard physical evidence of a crashed flying saucer, and, arguably, he was the first president to do something about it. It was his administration, from 1947 through 1952, that ordered the first reverse engineering project for the Roswell crash debris, his administration that ordered the interception of UFOs over Washington, DC, and the his was first administration that openly acknowledged the presence of flying saucers in American skies. President Truman even admitted publicly that conversations about flying saucers were held at the highest levels of the military and the government.[19]

All of the research conducted over the past seventy years indicated that the story of Roswell was not just a mass of confusion or some concocted story to explain the crash of a test aircraft, a Russian secret weapon, or a Japanese balloon bomb. It was a real event about the crash of something truly anomalous or otherworldly, the result of which was nothing less than a change in the course of human history. Despite all the rumors, the cover stories, the hype, and the debunking, out there in a high desert so lonely you'd think it was another world, human beings from an army airfield came into contact with entities from another place and time. And American history, in that moment, likely changed forever.

Chapter 9

WE LIKED IKE

Only weeks after the UFO invasion over Washington, DC, and months before he was nominated by the Republican Party for the 1952 presidential election, retired Supreme Allied Commander in Europe during World War II and retired General of the Army Dwight D. Eisenhower was aboard the aircraft carrier, the USS *Franklin D. Roosevelt* to witness NATO military exercises in the North Sea. The exercises, codenamed "Operation Mainbrace," were one of the first joint NATO, land, sea, and air exercises to test the abilities of the Allies to block Soviet submarines from entering the Atlantic Ocean in the advent of war. According to witnesses on the USS *FDR*, underwater UFOs, a bright light that was underwater, suddenly rose out of the water and flew off into the distance. It was an object no one onboard the *FDR* had ever seen before. At least one crew member on the bridge revealed, years later, Eisenhower himself saw the object and told sailors on the bridge that they should keep what they saw just between them and not talk about it. It was a secret. This might well have been Eisenhower's first eyewitness sighting of a UFO. But what Eisenhower saw that day would not be the last time he would come into contact with information, possibly including physical trace evidence, about visitations from ETs to planet Earth.

By the time Ike was president, the government's policy was to debunk UFO sightings publicly, even as officials became more secretive and maintained massive clandestine files on the subject.

The government's tactics had profound implications for virtually every aspect of unexplained phenomena research. The official attitude to dismiss UFO sightings as misguided or mistaken observations by the deluded extended to ESP, ghosts, apparitions, and a range of after-death communications, even when they occurred to First Families, thus relegating the paranormal to an elusive "netherworld" somewhere between fact and fiction, and supposedly far from the lofty heights of the White House, was government policy.

In *The Roswell Incident,* authors Charles Berlitz and William Moore claimed that in his first term, Eisenhower began to "make inquiries into the reality of the Roswell story."[1] But, incredibly, despite being a former army general, the authors contend, Ike did not have the clearances needed to obtain such highly classified intelligence, and never made public what he might have known about UFOs. There was even speculation that he was prepared at one point to address the American people on the subject, but demurred or was discouraged for reasons of national security. We know now that because of the existence of the MJ-12 and the Eisenhower Briefing Document and Ike's experience on the deck of the USS *FDR*, Eisenhower probably knew already about the nature of UFOs, their presence in our skies, and the government's attempts to intercept them.

Muroc

There are legends, some of them only rumors, describing a meeting between President Eisenhower and the extraterrestrials in February 1954, at Muroc airfield in California where he also saw their craft. But it wasn't just a meeting, it was the start of an alliance between President Eisenhower representing the United States and whatever culture the extraterrestrials represented. In one version of the story, Ike was supposed to be on a golfing vacation in Palm Springs, California, when he secretly slipped away to the airfield in Palmdale. According to the story, also in attendance at the meeting were Cardinal James McIntyre, the Bishop of Los Angeles, who was a close confidant of Pope Pius XII, and a psychic and mystic named Gerald Light, who President Eisenhower believed would help him communicate with the ETs, perhaps telepathically. Gordon Light was known as a self-published, self-professed authority on talking to extraterrestrial beings. And he wrote a letter to a friend describing this meeting, which reads:

Gerald Light
10545 Scenario Lane Los Angeles, California
Mr. Meade Layne San Diego, California

My dear Friend:

I have just returned from Muroc [later renamed Edwards Air Force Base]. The report is true—devastatingly true!

I made the journey in company with Franklin Allen of the Hearst papers and Edwin Nourse of Brookings Institute (Truman's erstwhile financial advisor) and Bishop MacIntyre of L.A. (confidential names for the present, please).

When we were allowed to enter the restricted section (after about six hours in which we were checked on every possible item, event, incident and aspect of our personal and public lives), I had the distinct feeling that the world had come to an end with fantastic realism. For I have never seen so many human beings in a state of complete collapse and confusion, as they realized that their own world had indeed ended with such finality as to beggar description. The reality of the "other plane" aero forms is now and forever removed from the realms of speculation and made a rather painful part of the consciousness of every responsible scientific and political group.

During my two days' visit I saw five separate and distinct types of aircraft being studied and handled by our Air Force officials—with the assistance and permission of the Etherians! I have no words to express my reactions.

It has finally happened. It is now a matter of history.

President Eisenhower, as you may already know, was spirited over to Muroc one night during his visit to Palm Springs recently. And it is my conviction that he will ignore the terrific conflict between the various "authorities" and go directly to the people via radio and television—if the impasse continues much longer. From what I could gather, an official statement to the country is being prepared for delivery about the middle of May.

I will leave it to your own excellent powers of deduction to construct a fitting picture of the mental and emotional pandemonium that is now shattering the consciousness of hundreds of our scientific "authorities" and all the pundits of the various specialized knowledge that make up our current physics. In some instance I could not stifle a wave of pity that arose in my own being as I watched the pathetic bewilderment of rather brilliant brains struggling to make some sort of rational explanation which would enable them to retain their familiar theories and concepts. And I thanked my own destiny for having long ago pushed me into the metaphysical woods and compelled me to find my way out. To watch strong minds cringe before totally irreconcilable aspects of "science" is not a pleasant thing. I had forgotten how commonplace things as dematerialization of "solid" objects had become to my own mind. The coming and going of an etheric, or spirit, body has been so familiar to me these many years I had forgotten that such a manifestation could snap the mental balance of a man not so conditioned. I shall never forget those forty-eight hours at Muroc!

G.L. [2]

Among others at the meeting along with Cardinal McIntyre were Dr. Edward Nourse, the chief economic advisor to President Truman, and a retired reporter for the Hearst newspaper chain, Frank Allen.[3] As outrageous as this story seems, facts do support the story that President Eisenhower did have a missing night when it seemed that he was completely off the grid after he was supposedly secreted away to Palmdale.

Among UFO historians, there are a number of theories, none of which are supported by hard evidence, regarding the rationale for Ike's purported meeting with the ETs, not the least of which were the many incursions of US airspace since 1945 including the DC incursion and the loss of American fighter aircraft pursuing the flying saucers. One theory posits that Eisenhower, after having received the briefing documents and having seen a flying saucer with his own eyes during Operation Mainbrace, sought to find out what these visitors wanted, whether they were hostile or otherwise posed a threat to US national security, or had any technology that would be beneficial to humanity.

One curious aspect to the meeting was a story about Eisenhower's striking a deal with the ETs, not a treaty, which, constitutionally, would have required Senate ratification, but an "open skies" agreement that acknowledged their presence around our planet in exchange for a non-aggression understanding and a transfer of their technology to the United States. Under that agreement, UFO researchers have suggested, was an understanding that because the ETs were interested in researching our species, Ike made a backdoor acquiescence that humans might be abducted and studied as long as the ETs agreed that no physical harm would come to them. The open skies agreement that Ike struck with the ETs might have been the basis for the open skies agreement that Eisenhower struck with Nikita Khrushchev years later. Another strange story coming out of this Muroc meeting regarded the very nature of the UFO landing itself as the basis for the dramatic UFO landing scene in the Steven Spielberg film *Close Encounters of the Third Kind.*

The RB47 Incident

Imagine that you're the president of the United States, sitting in the Oval Office after a brilliant military career that was capped off by your stunning victory over one of the most powerful war machines in history. Now imagine that you are facing an enemy across both oceans whose leader has vowed that your government will fall and a revolution will destroy the Constitution, which you have sworn to preserve and protect. Now imagine further that you, with your own eyes, witnessed a craft whose origins you could not explain rise out of the water, undetected at first, near the hull of your navy's most advanced aircraft carrier on a mission to block the Soviets from entering the Atlantic via the North Sea. Imagine all that and then imagine what it must have been like when your air force's most advanced electronic countermeasures aircraft is completely bamboozled by an anomalous craft that, literally, flies circles around it and can disappear from its radar at will. Imagine all of that, assuming that the event is real and not a screw-up by the best trained crew in the air force, and you will come to one of two startling conclusions: either the Soviets have a new weapon that is a Cold War game changer, or, worse, we've been intercepted or invaded by extraterrestrials. How does a president deal with a situation like this?

This incident in question took place early in Ike's second term, July 17, 1957 at 4 a.m. local time, when the air force's most advanced electronic surveillance and countermeasures electronic warfare aircraft, the RB47, flying on a training mission over the Gulf of Mexico, near the city of Gulfport, Mississippi, picked up a strange target on radar that kept moving around the plane in ways that the crew could not easily explain. As the plane turned west, heading for Texas, the pilots also saw a strange blue pulsating light in front of the plane, whose origin they could not explain. At the same time, on the number 2 radar monitor, the radar officer, or "Raven," detected a signal he could not explain. For the next ninety minutes and 700 miles, the aircraft's electronic intelligence apparatus, ground radar installations, and even the pilots in the cockpit observed what many ufologists believe was a strange object tracking the RB47.

The signal on one of the radar scopes picked up a target on the right side of the aircraft, which signal quickly climbed upscope—meaning that it was moving along the right side of the RB47 towards the nose, then even as it intermittently disappeared from the scope, it crossed the front of the plane and showed up on the scope covering the left side of the plane and seemed to climb down the scope. It seemed to be tracking the aircraft as it flew northwest. Because what the crew picked up on their radar was also picked up at the same time by ground radar, confirming the position of the object relative to the RB47, it thus made this incident a multiply witnessed event.

The RB47 electronic surveillance and countermeasures aircraft either was followed by a UFO, which the crew said they visually tracked as well as tracked on their radar, or whether what they saw and picked up was an American Airlines commercial flight, which is what the Project Blue Book case managers reported. Skeptics such as the late aviation journalist and CIA embed Phil Klass argued that the electronic counter warfare officers on the RB47 actually picked up ground radars and mistook them for another aircraft. The Condon Report, released ten years later, said this case was unexplained. University of Arizona physicist Professor James McDonald believed this was a real UFO case in which an anomalous object, capable of obscuring itself from our radar while changing position instantly, presented a viable threat to our national security because it was able to out maneuver our most advanced electronic countermeasures aircraft while concealing itself from the aircraft's radar.

The issue, in order to oversimplify a complicated case of conflicting interpretations of electronic and visual data, circles around whether a well-trained crew of electronic warfare countermeasures personnel could distinguish between the pattern of ground radar stations transmitting a signal that the ECM (radar) officers, also called "Ravens" or "Crows," said was following them in an unconventional manner winked out, crossed the plane's flight path, reappeared on the other side, then followed them across a few states, and then disappeared. The highly technical and well-researched investigations are persuasive. However, even the most technical analysis of the plane's flight path, capabilities, functionality of the ground radar and the atmospheric conditions that could have lengthened their signals, as well as the actual object that the pilot and copilot visually observed, do have a level of conjecture to them that is less than conclusive.

Former NSA officer Bill Scott, retired Rocky Mountain Bureau Chief for *Aviation Week* and the co-author along with retired naval officer Mike Coumatos and Bill Birnes of *Space Wars* and *Counter Space,*[4] knew the specifications and mission of the RB47, especially its electronic warfare countermeasures capability, and what the ECM officers were tasked to do in the "bubble," the radar pod in the RB47's bomb bay. The RB47 was the backbone of the US Air Force surveillance mission. It was specifically outfitted not to carry weaponry, but to carry sophisticated radar and analysis equipment for electronic intelligence-gathering. It was the functionality of this equipment and the analytical abilities of the Ravens that lay at the heart of what the crew actually had seen and experienced.

The RB47 was the reconnaissance backbone of the USAF, specifically configured for recon and surveillance, scrubbing for electronic intelligence such as the capabilities of enemy radar and tracking targets. Therefore, when the crew picked up something they thought might be anomalous, it caught the interest of the commander in chief, President Eisenhower, because of the real data collected by the instruments onboard that allowed the air force to evaluate the nature of the event. As Bill Scott, during his appearance on the History Channel's *UFO Hunters,* said about the RB47 missions, "the plane would fly along the borders of the Soviet Union and listen for the radars transmitting along the ground. We're talking about technology in the 1950s, but the RB47 was the state of the art and critical to our national security."

Bill Scott continued, "At the time, this was our frontline fighting force, the Strategic Air Command. These elint [electronic intelligence] operators would have been very well trained because the fate of the country was in their hands and in the hands of the pilots who flew these airplanes." According to Scott, the crew had at their disposal "tried and true hardware of the time. So it was cutting edge, but very well maintained."

Despite the RB47's being the most advanced equipment in service, the officer who first spotted the object on the electrical intelligence boards thought initially that the equipment might have been malfunctioning. But, as is typical among military hardware specialists, they had one of the other operators double-check the equipment, and he tried to reestablish the signal's reception on a different set of antennas on the RB47. Thus, the crew substantiated very early on that the equipment was functioning perfectly according to specifications. In other words, there was no perceivable malfunction.

In a short while, however, the crew was about to get visual confirmation of what they had seen on their scopes just as the RB47 headed toward Jackson, Mississippi, when one of the pilots spotted what he thought were the landing lights of another jet coming in fast. But these were no landing lights. As the single blue light closed rapidly on the RB47, the pilot, Colonel Lewis D. Chase, alerted the crew that they should prepare themselves for sudden evasive maneuvers. But before he could do anything, he and his copilot saw the light instantaneously change direction right in front of them and flash across their flight path. Then it blinked out. Professor James McDonald interviewed, and recorded, the cockpit crew, obtaining firsthand eyewitness confirmation of what they saw.

"I'm looking at it as a pilot at night," Colonel Chase told McDonald. "Seeing a light source that I interpreted as an airplane. But then suddenly it's flying in front of me so fast that I couldn't respond."

"Did it suddenly blink out or did it take off at an appreciable velocity?" McDonald asked Chase.

"No, just gone," Chase responded. "Just disappeared."

"That was quite an intercept you were flying there," McDonald suggested. "That is really very strange."

In the McDonald report, Colonel Chase claimed that the object moved at a velocity that "he'd never seen matched in his flight experience."

But some skeptics and debunkers have argued that the cockpit crew only had a sighting of a falling meteor that crossed their path and then disappeared, which, they conjectured, would explain the velocity and the object's blinking out. Huh? Meteorites don't change direction in flight. They don't flash on and then off. Instead they flame out. This object didn't.

This was our most advanced electronic intelligence and counter warfare in our arsenal, and something played cat and mouse with it while the crew simply checked and rechecked their apparatus. Now ask yourself whether a well-trained veteran crew, with a mix of expertise and technology, would have not recognized the playing of ground radars around the RB47's antennas and a meteor falling through the sky. And, by the way, did anyone else see that meteor? Did the falling meteor show up on the ground radars that the RB47 was supposedly picking up? And did the meteor simply burn up or did it hit the ground somewhere? All things to be considered if there were a conventional explanation for this incident, a conventional explanation that one of our country's most important climate physicists could not figure out and had to ask why were there so many "ifs" that defy the measure of plausibility?

Bill Scott was adamant that in 1957, the year of the incident, "you didn't have an aircraft that could jump from 11 o'clock to 2 o'clock instantaneously and then just blink out." Assuming, now for the sake of argument, that the target on the RB47's Ravens' radar scope was a craft and not the plane's radar picking up radar signals from ground radar emplacements, why would the craft, and not a meteorite, be following them as the aircraft crossed the Gulf coast, then turned west towards Dallas and that's when the pilot had a visual sighting and began taking evasive action?

"They started accelerating and decelerating to see if they could lose the object," Bill Scott said. As the pilots watched the object in their path, the radar operators at Duncanville said they also had an anomalous target on their screens that corresponded to the position of the object the pilots were seeing. Then the light blinked out, Bill Scott said, "when the pilots reported that the light disappeared, the ground-based radar at Duncanville said they lost the object as well and the elint operators on the RB47, the Ravens, said they lost the target from their scopes at the same time."

This correspondence between the visual sighting and the radar sightings is significant. But things got even more interesting when the object reappeared visually and, again, turned up on the Duncanville radar and the elint scopes on the RB47. Everything associated with the appearance of this target, the visual sighting, the ground-based radar hit, and the elint tracking on the plane's scopes were all seeing this light/target at the same time and confirming each other's tracking the target, thus confirming that they were seeing the same thing visually as well as electronically.

After the crew had made their turn towards Texas, heading for their home base, Colonel Chase requested and received permission to pursue the target that appeared on their scopes. The object, they confirmed visually, was at a lower altitude so the pilot dived towards the object and suddenly, the object stopped right in midair. Does the sound like a meteorite? The plane overshot the object and then the crew started a left-hand turn to get back to where they could see the object, at which point they were running low on fuel and decided to break off the pursuit and head back for home. But this time, the object pursued them again.

"So they took off to the north and headed back to Forbes Air Force Base in Kansas, and that's when the object stayed with them," Bill Scott said. And the object stayed with the RB47 almost half the way back to Forbes, disappearing from the radar tracking over Oklahoma City. The ninety-minute chase of an object that both tracked and evaded the RB47, bouncing from one radar scope to another and turning up on ground radar, left the plane's crew and ground-based personnel dumbfounded. None of them had an explanation for what had just occurred. All of these maneuvers were impossible for an aircraft in 1957, and that's what troubled the former Supreme Allied Commander, Dwight Eisenhower, who had also read about the foo fighters intercepting American fighter planes over Europe.

"In 1957, you don't have an aircraft that can just blink out both in the electromagnetic spectrum as well as in the visual spectrum," Bill said. "And I would say that we didn't have anything back then and don't have anything like that now."

However, despite all of these anomalies and events that defied conventional explanation, Project Blue Book dismissed the event out of hand even though it involved the air force's most advanced electronic counter warfare weapons system. Blue Book said that the event was really a misidentification of the near collision between two commercial DC6 airliners near Salt Flat, Texas. But they gave no explanation for their conclusion except to say that "it was definitely established by the CAA that the object observed in the vicinity of Dallas and Fort Worth was an airliner." However, Blue Book's conclusion is as implausible as it is unlikely. The RB47's flight path didn't come within 400 miles of Salt Flat, Texas, where Blue Book said the DC6 airliners were. And Professor McDonald used the reference in Blue Book to comment on the air force's poor methods of investigation. We now suspect, of course, that the public Blue Book was simply a way to satisfy the public with conventional explanations for unexplainable UFO events. The RB47 incident? Blue Book categorized it as "Identified."

Professor James McDonald never agreed with the Blue Book conclusion because nothing they said comported with the crew's own observations, and the case remained "Unidentified."

President Eisenhower was a transformational president not specifically for what he did, but for what he kept the world from doing. He came into office on the wings of one of the greatest military careers in American history at a time when two adversaries, the US and the USSR, possessed nuclear weapons that could have started the extinction of life on earth in as little as seventy-two hours from launch. Yet he forged agreements with our enemies that kept, in JFK's words, a "bitter peace," but a peace nevertheless. He may not have been one of our most charismatic leaders, but in a world of itchy trigger fingers, he protected his country and protected his planet. And after almost sixty years, it's nice to know.

Chapter 10

JFK, THE MARILYN MONROE CONSPIRACY, AND ALL THE PRESIDENT'S UFO MEMOS

The Marilyn UFO Memo

Marilyn Monroe had a crush on John F. Kennedy. She thought he was different from the other guys she knew. Joe DiMaggio was too controlling. He wanted her to leave show business. Arthur Miller, too cerebral, and she felt constrained. But Jack Kennedy dazzled her and because he gave her the attention she craved, she fell for him like a child. But it would go nowhere, and Marilyn began to act out. Frank Sinatra and Dean Martin, both friends of the Kennedys, would take her up to Frank's Cal Neva lodge in Tahoe to dry her out. Peter Lawford became her minder for the Kennedys, carrying tales back and forth like Iago, which is what movie star Ava Gardner once called him when he was carrying back to her stories of her husband Mickey Rooney's infidelity with underage girls.

Then Jack Kennedy refused to take Marilyn's calls and it was Bobby Kennedy's turn to romance her. Which he did. And when he stopped taking her calls, Marilyn got angry. And when she got angry, she got out of control and she began threatening the Kennedys, threatening to go public about her affair with Jack, threatening to talk about their both seeing Max "Dr. Feelgood" Jacobson and taking his special vitamin cocktail—actually forty milligrams of methamphetamine—that he injected directly into their thighs.[1] Marilyn was seething and was talking out of school. And that's why her phone was tapped by both the FBI, because Hoover was looking for anything he could get on his archenemy Bobby Kennedy, and by the CIA, who had a deep cover operative in the White House and in the entertainment/photography industry, Mark Shaw.[2] The CIA and the FBI knew that Kennedy was receiving methamphetamine injections from Max Jacobson and they knew that Jacobson in the 1930s, after he fled Berlin with his family and turned up in Vienna, had been turned into an operative by Soviet military intelligence. He was dangerous. And now so was Marilyn. Frank Sinatra and Dean Martin had taken Marilyn up to Cal Neva in Tahoe to dry her out, but she started drinking heavily when she got back to Beverly Hills. And Marilyn's acting out came in the wake of her failed feature film *The Misfits*, written by Arthur Miller and co-starring Clark Gable. Then, when she was fired from the film *Something's Got To Give* with co-star Dean Martin—itself a remake of Garson Kanin's *My Favorite Wife* with Cary Grant and Irene Dunne, and a film canceled and then later remade with Doris Day, James Garner, and Polly Bergen called *Move Over Darling*—she sank into a deep depression. A depressed and angry Marilyn became out of control and dangerous to the Kennedys. Now it was left to actor Peter Lawford, JFK's and Bobby's brother-in-law, to mind her.

Just days before her death, ruled a suicide by the LA County Medical Examiner even though serious questions remain even fifty-five years later particularly because of a tell-tale raised bruise on her right thigh, Marilyn made a phone call to Attorney General Robert Kennedy at the US Department of Justice. Because Marilyn was under surveillance because of her relationship to the president, and under surveillance from the FBI because of her relationship to Robert Kennedy, her threatening phone call was wiretapped, recorded, and a transcript made, with the appropriate file designations, which transcript historian and researcher Dr. Donald Burleson printed in his book, *UFOs and the Murder of Marilyn Monroe*.[3]

In her phone call, Marilyn made reference to a "secret diary," her journal, she said, that if released would shock the nation because of its disclosures. In the CIA transcript, dated August 3, 1962,[4] two days before Marilyn's officially designated suicide, although it was probably a murder made to look like a suicide, it is clear that Marilyn was in possession of details that could only have been passed on to her by President Kennedy regarding some of the most closely held secrets of the Kennedy administration. For example, Burleson points out that one of the revelations of the transcript was "the visit by the President at a secret air base for the purpose of inspecting things from outer space."[5] Was this secret air base Area 51? What were the "things from outer space," parts of a UFO, a downed Soviet satellite, or maybe only a meteorite? We don't know, but according to UFO researchers, the thinking is that there were either alien spacecraft or parts of UFOs that the government was trying to reverse engineer that were stored at the base. But the transcript became even more incredible when Marilyn threatened to talk about the CIA plans to kill Castro and the fact that Marilyn had discussed things like "little men from space" with columnist Dorothy Kilgallen, who died shortly after she had interviewed Jack Ruby, Lee Harvey Oswald's killer, in prison. One can, of course, question the provenance of the CIA memo, but there is a clue to its veracity in the file designations ordering the disposition of the document to files named "Moon Dust," dealing with information about space-based phenomena, specifically UFOs, and "Blue Fly," the retrieval of objects from outer space, again, think UFOs.[6]

From the CIA's perspective, the president was completely hooked on methamphetamine, which was seriously impairing his ability to govern right up to the point where the US was literally minutes away from absorbing the launch of a ballistic missile from a Soviet submarine off the coast of Cuba. If this weren't bad enough, he was sharing secrets of UFOs and ETs with a woman with whom he was having a torrid affair, who, herself, was in the public eye and whose disclosures would rock the government. At the same time, the president was experiencing drug-induced nervous breakdowns—drugs administered by Dr. Max Jacobson, a Soviet intelligence operative—which were threatening to become public in New York's Carlyle Hotel and required the president to be placed in a jury-rigged straitjacket to control him from running naked through the halls of the Carlyle. JFK's near-breakdown took place with the White House and New York press gathered downstairs at the hotel and required the services of the Chief of Psychiatry at Cornell/Downstate, Dr. Lawrence Hatterer, to inject him with a heavy dose of phenobarbital to sedate him.[7] JFK was spilling state secrets, was putting himself and the presidency in jeopardy by slipping away from his protection to dally with call girls, and had almost brought the world to a nuclear war. At a time before the ratification of the Twenty-Fifth amendment, he had to go: one way or another.

From the FBI's perspective, things were not much better. In fact, J. Edgar Hoover believed, things could be even worse. His agents told him that Marilyn Monroe was not only out of control, a methamphetamine addict at JFK's Madison Square Garden birthday party where she sang "Happy Birthday, Mr. President" while under the influence of a heavy dose of Max Jacobson's drugs, but that she was a danger to the administration. He also knew from internal memos that Marilyn was suicidal,[8] a fact also known to the Kennedys and Marilyn's friend, JFK's brother-in-law and Marilyn's minder, actor Peter Lawford. He, of course, would know about what a minder was supposed to do because he was good friends with newly-weds Mickey Rooney and Ava Gardner and knew that Mickey and Ava had a minder from the very studio where all three of them worked, MGM, and knew the official minder there, Les Peterson.[9]

Of course, the FBI would have been investigating the ties among Lawford's friends in Frank Sinatra's Rat Pack because of the links between the celebrities and mob boss Sam Giancana as well as Joe Kennedy's links to the mob dating back to Prohibition when Papa Joe was dealing with Hoover's old friend Meyer Lansky, whom he once called "the best businessman I know" when Joe Kennedy and Lansky were smuggling liquor into New York harbor.[10] Hoover was also friends with New York mob boss Frank Costello.[11] One of the early memos to Hoover revealed that on the date of her death, Marilyn's housekeeper put two bottles of pills, probably Nembutal, on Marilyn's night table for the purpose, the memo continues, "to induce suicide."[12]

Marilyn had been through another shock before she died after director George Cukor fired her from *Something's Got To Give*. Although in trouble in Hollywood after the failure of *The Misfits*, Dean Martin and Frank Sinatra convinced Cukor to give her a shot. But Marilyn was very depressed and was acting up on the set, drinking, and coming in very late for her calls. Finally an exasperated Cukor fired her and then the studio canceled the movie.

A year after Marilyn's death, an FBI memo asserts that her "killer" is still at large and identifies him, not by name, but by saying that he's a married man, a celebrity well known to the public and "the world over," and a father.[13] The memo identifies three individuals who knew that Marilyn at the time of her death was having an affair with Bobby Kennedy, who, the memo says, had promised to leave his wife, Ethel, even though Marilyn realized that Bobby was probably lying to her. Hence the phone call to Bobby threatening both him and his brother and threatening to release her secret diary. But who were the three people that the internal FBI memo cites as being present on the day of her death and who might not be above suspicion in a conspiracy to set Marilyn's suicide? They were Peter Lawford, Marilyn's housekeeper Eunice Murray, and her publicist Pat Newcombe, who, the writer of the FBI memo alleges, made sure Marilyn had "access to lots of sleeping pills and other barbiturates to try to 'induce suicide'." Peter Lawford, the memo says, knew about Marilyn's ideations of suicide and knew that providing her the access to the means of suicide might get her out of the way and end her threat to the Kennedys. But neither the FBI nor the CIA knew of another person who was present on the day of her death, someone other than her psychiatrist Ralph Greenson, but a physician upon whom Marilyn not only relied, but whom she trusted implicitly because he had become the sole source of her physical and mental well-being, even if only for a few hours each session. It was Dr. Feelgood's son, Dr. Thomas Jacobson, who was handling his father's patients in Los Angeles and treating Marilyn.[14]

In interviews with the authors of *Dr. Feelgood*, it was revealed that Max Jacobson's son was at Marilyn's house before her death where he had dropped off vials of methamphetamine for her to inject, which she did. Given her propensity for suicide and the clinically depressive effects of methamphetamine when it wore off, it took very little for a mentally distraught Marilyn Monroe, feeling rejected by both Jack and Bobby Kennedy, feeling used and abandoned by the Kennedys, and more than hopelessly disconsolate over the failure of her two latest motion pictures, to have overdosed on meth and then taken overdoses of Nembutal, lapsed into a coma, and left to sink into death by Peter Lawford, who shooed away any attempts to awaken her. Years after Marilyn's death, Peter Lawford revealed his complicity and his feelings of guilt to one of his former managers.

Army R&D in the Days After Roswell

Retired United States Army lieutenant colonel Philip J. Corso, a former intelligence officer in World War II, who, at the beginning of the Kennedy administration in 1961, had served out his final tenure in the army reserve as the Inspector General of the Maryland National Guard, was summoned to the Pentagon by Lieutenant General Arthur Trudeau, the commandant of the army's Office of Research and Development, to oversee the Foreign Technology Desk. But, as Trudeau, one of the heroes of the Korean War, explained to Corso, who had stood up for him when he was on the national security staff at the Eisenhower White House after Trudeau had revealed that the East German spy ring, the Stasi, had penetrated Gehlen's CIA-affiliated spying operation right under the nose of Allen Dulles, he needed him to cover his back, be a messenger delivering UFO technology from the crash at Roswell and budget funding to national defense contractors for the reverse engineering of that technology. On Kennedy's watch, the army systematically, with the support of Senator Strom Thurmond, selectively distributed bits and pieces of Roswell alien technology to companies such as AT&T, the army laser laboratory at Columbia University, and Fort Belvoir for the development of night vision goggles. While the army built up its brain trust, which included the famous German rocket scientist Hermann Oberth, the father of the Nazis' missile program and the person who helped construct the German flying saucer, who had stunningly revealed about Nazi technology that the Germans had been helped "by people from other worlds," Corso delivered integrated circuitry to computer manufacturers, fiber optic flex cable to Bell Labs, and the outer covering of an ET to Monsanto. Perhaps this was the reason that JFK was so aware of the presence of ET technology in government files and, at the beginning of America's space program, sought to harvest that UFO technology.

The Betty and Barney Hill Abductions

The brand-new Kennedy administration was in shock after the failure of the Bay of Pigs invasion in April 1961, less than three full months after JFK's inauguration. Then three months after the debacle on the beaches of Cuba, an over-drugged President Kennedy, stinging from his first defeat the hands of Fidel Castro, had to face off against Nikita Khrushchev in Vienna. Kennedy's Dr. Feelgood, who accompanied the president, was waiting in an anteroom next door administering shot after shot of meth-amphetamine to keep the president upright, where he, in his own words, experienced "the worst day of his life" after the Soviet leader threatened to wall off Berlin and prevent the allies from entering the city.[15] These were both political and national security events. But just two months after

Kennedy's failure in Vienna, he had to face another crisis, this one conveniently tucked under the rug until the national news got ahold of it. After the news broke, Kennedy was staring at the national cover story of a married couple who claimed, during psychiatric therapy, that they had been abducted by extraterrestrials along a lonely road in New Hampshire. The story, evolving from 1961 to 1962, resonated across the country and got the air force involved, involvement that would surely have reached the desk of the commander in chief.

Betty and Barney Hill, in a racially mixed marriage years before the US Supreme Court ruling striking down laws banning interracial marriages,[16] returning from their planned their vacation in Canada, were driving home to Portsmouth, New Hampshire, along a lonely country road in 1961 when a light in the sky they could not identify, identified them, changing their lives in ways they could not imagine, and their resulting story awakened Mr. and Mrs. America to the fact that we are not alone in the universe.

Barney and Betty were a very private couple partly because spouses in a mixed-race marriage were uncommon enough in New Hampshire or anywhere else in 1961. They both worked for social causes in addition to their regular jobs. Barney worked for the United States Post Office and sat on the local Civil Rights Commission. Betty was a social worker, and both of them were members of the NAACP. At the dawn of the 1960s, a time before the civil rights movement began in earnest under President Lyndon Johnson, Betty and Barney were advocates of social change. But in the tiny towns of America's conservative New England, the last thing the Hills wanted was national publicity, which was exactly what they got—the cover of *Look* magazine—when they had the misfortune to spot a huge bright light up in the night sky following their car along a country road outside of Groveton, New Hampshire, on September 19, 1961.

Barney suffered from hypertension at the time and also suffered from the symptoms of a stomach ulcer. Maybe it was the pressure from his job at the post office. Maybe it was the rising number of civil rights clashes in the South that were dominating the television news. Whatever it was, the Hills needed a break and took a quick vacation, getting away from it all, in Quebec. On the night of September 19, they were driving home from Quebec to Portsmouth along Route 3 in the utter darkness when, right around midnight, a very bright light way off in the distance caught their attention. It could have been a star, but it was unlike any star they had seen before. Betty believed that it was one of the satellites, maybe Telstar, that the US had just put up. Barney, however, thought that it was a plane that seemed unusually bright against the very dark night sky. He was worried and said during a hypnotic regression session with Dr. Benjamin Simon long after the event, "I first saw it in front of the car but then it swung around to the rear."

The light seemed brighter than a conventional aircraft and it seemed to be tracking them. Just the thought that there was a plane following the route they were taking made Barney nervous. He said, "Look, it's following us, Betty." And told Dr. Simon during a later therapy session that he was "hurrying to get away."[17]

As they drove, Barney constantly looked out his window to keep track of the light, which seemed to be getting closer. Betty and Barney kept talking about what that light might be, Barney now believing he was looking at a disk-shaped object. He made a turn, later telling Dr. Simon that he didn't know why he had to make that turn. Then, he realized that he was lost.

The light kept following them and seemed to be positioning itself right above the car. Finally, at Betty's urging, because she wanted to make sure of what the object was, Barney stopped the car at the Mount Cleveland picnic area and the couple got out. Once outside of the car, Betty pulled out a pair of binoculars and got a better look at the object against the light of the moon. It wasn't a satellite at all, she realized, but something else, a circular object that was illuminated with multi-colored lights. The object crossed the face of the moon and then swooped down. It was rotating and flashing a blue-white light. It passed them heading north towards Vermont, and Betty could see it fly over the tramway over the Franconia Notch.

The Hills got back and the car and started driving again, heading for the safety of the nearest town. Whatever the thing was, Betty told her niece, Kathleen Marden, the following morning, she and Barney had the feeling that the object had taken notice of them.[18] They were nervous, all alone on that New England country road at night, with nothing between them and that strange object. They soon found that what initially had been a nervous reaction to something anomalous in the sky became downright frightful as the object suddenly shifted its direction and now seemed to be tracking their car more closely, staying low to the horizon and matching their speed.

The object seemed to follow them for a few more miles and then suddenly began a descent toward their car. When the object stopped, hovering about a hundred feet above the car, directly in front of the windshield, Barney stopped the car in the middle of the road, slammed it into park, and stared up at the object. Through what looked to him to be a row of windows, Barney could see eight to ten actual figures, not people, but humanoid shapes, and they were looking down at him. Barney wanted out of there, and fast. As he told Dr. Simon, "I was driving and driving, and I made a turn."[19] He made a left-hand turn onto a dirt road.

Ahead of him he saw what he believed to be six to eight men in the road. Barney was confused. What were these men doing in the middle of the road after midnight? It dawned on Barney and Betty as soon as what they thought were men broke into two groups and began to approach them that, to their shock, these weren't men at all.

Barney reached for the pistol he had been carrying, but it was too late. The figures approached the doors and forced them open. While Barney struggled to run away, Betty also tried to escape. As she told her niece Kathy Marden over the phone the next day, she remembered trying to flee the car through her door but before she could get any distance, she was immediately grabbed by the small humanoids and dragged, against her will, to a small clearing in the woods about a hundred and fifty feet away.[20] In the meantime, the humanoids had already grabbed Barney, preventing his escape, and dragged him towards the clearing, too. Betty told her niece that she remembered the entire incident consciously, even seeing Barney next to her in the clearing. She remembered struggling with the humanoids, who tore her dress, and then she saw a ramp descend from something above them. And that was all she remembered until she and Barney found themselves driving along a very familiar road coming up to their driveway. It was almost dawn, and they had no idea where they had been for the previous few hours.

Later that day, Betty phoned her sister and talked to her niece Kathy about the incident. She described the craft, talked about the small humanoids, who grabbed her, said being dragged into the clearing with Barney was the last thing she remembered before returning to awareness just as they pulled up on the road near their home. She remembered nothing in between, except for the disconcerting feeling that in the hours that were missing from her life something had happened to her after she struggled with those creatures. All she knew at that point was that there was a strange pink powder residue on the dress she had been wearing, and the dress was torn. That pink residue resonated with a missing part of her memory, telling her that something very traumatic had happened.

This would be the beginning of a very revealing inner journey for Betty, as well as for Barney, because try as they might, the Hills simply could not return to a normal life. Betty said in interviews years later that Barney, who had suffered from hypertension—a disease that finally killed him—continued to be uneasy and complained of all sorts of physical problems. In particular, Betty said, Barney was unusually worried about his genitals and complained of pains in that region. He said his stomach was bothering him, and he was having trouble sleeping at night. Moreover, try as he might to calm down, Barney seemed to be unusually irritable, as if something were eating at him that he couldn't resolve. Thinking that this was some kind of underlying physical condition, Betty took him to their family doctor. But after a series of exams, the doctor could find nothing wrong with Barney except that his blood pressure was elevated.

While Barney was manifesting physical symptoms, Betty was having serious nightmares, repressed memories of what had happened to her after the humanoid creatures had taken her. She kept the dress she was wearing that night, a dress that had pink stains where she remembered the humanoids had touched her. She never had that dress cleaned because she believed she had an actual piece of physical evidence from what had happened to her. But beyond that she had only her conscious memories and her dreams.

Betty Hill also reported the incident to Pease Air Force Base, the local base in that area of New Hampshire. An investigator from Pease called her back, interviewed her, and then reported that Betty had probably seen the planet Jupiter. Betty had left out many important details of her encounter, she said years later, because she was afraid of being thought insane. The Hill incident became part of the air force's Project Blue Book. Air force officials, however, knew that there had been an unidentified object in the area that night from other reports, but in keeping with their protocols, they denied anything to do with a UFO.

Barney Hill's condition continued to deteriorate. His doctor, after hearing repeated complaints of physical ailments from his patient and after finding nothing physically wrong with him that he could correlate to Barney's symptoms, finally suggested that the problem might be emotional. Emotional problems, he advised, do, many times, create the impression of physical symptoms even though there might be no underlying physical causes. What Barney's doctor told him was also reinforced by Air Force Captain Ben Swett, who talked about hypnosis at the Hills' church in November 1962. Although the Hills asked Swett to hypnotize them to help them remember, Swett declined, saying he was not qualified. But he advised them to speak to their doctor so as to find a qualified therapist. Barney brought this up to his doctor.

Upon his doctor's advice, Barney visited a psychiatrist, who, after a session with him in which he learned of Barney's loss of memory, referred him to Dr. Benjamin Simon. Dr. Simon had played an important role in World War II, helping pilots suffering from hysterical symptoms to recover lost memories so as to confront and integrate the underlying trauma.

During the two years between their encounter and their visits to Dr. Benjamin Simon, Betty was also suffering from nightmares, dreams that played back a version of the events that had taken place aboard the ship. She dreamed about communicating with her abductors, now referring to them as otherworldly, dreamed about the medical experiments, and dreamed about being taken aboard the craft. These nightmares were persistent and would ultimately play a role in Dr. Simon's opinions about what might have been the origin of the stories the Hills told under hypnosis.

Dr. Simon regressed both Barney and Betty during separate treatment sessions, recording their responses on tape and asking his assistant to transcribe the recordings. Upon the conclusion of each session, he instructed both Betty and Barney not to remember what they told him, believing they were not ready to integrate what they told him under hypnosis into their conscious memories until they were prepared to do so in follow-up sessions. Kathleen Marden and Stanton Friedman in *Captured* write that Barney's emotional reactions during his regression sessions were so intense that Dr. Simon had to stop them at points because he was afraid of Barney's high blood pressure problems.

As the sessions progressed, the stories Barney and Betty told were nevertheless astounding. Separately, each told the same story of being placed into a trance-like state by the humanoid creatures, taken from the clearing by force onto what Barney and Betty described as a craft of some sort, placed under restraint on examination tables, and then probed and examined by creatures they described as "aliens" from another world. The details of the incident they both reported were so precise that Dr. Simon was shocked by these revelations, accounts he believed that his patients believed to be true even if he couldn't lend his professional credibility to the actuality of extra-terrestrials abducting human beings.

Among the elements of the Hills' stories that rang credible for Dr. Simon—at least insofar as their patients' joint and separate beliefs that the incident had actually taken place—were the correlations of the descriptions of the events from two different perspectives that supported each other. He believed from his initial examinations of the Hills that they had experienced true memory loss, a form of amnesia that seemed to have resulted from some sort of trauma. He heard from Betty's sessions that the amnesia was induced by her belief in something that their abductors did to them. They had been instructed, commanded, to forget everything and restart their memories only when they realized they were driving home at dawn. Yet, when the Hills told their separate stories under hypnosis, they both described being placed into a twilight state, taken aboard what seemed like the same craft, and made the subject of painful invasive medical-type examinations.

Barney, in particular, described an examination of his genitals, embarrassing as well as painful, that resulted in his being overly concerned about his genitals in his post-traumatic waking state. Physical examinations of Barney's genitals also showed that a pattern of warts had developed around them, a pattern which was consistent with his description of the place where his examiners placed their examination instrument. This was one piece of corroborative evidence which many people overlooked, but which was, in fact, extrinsic physical evidence that something had happened to Barney.

For her part, Betty related that her exchanges with entities she described as extraterrestrials were more cordial than Barney's. In one instance, an examiner tried to pull out her teeth, but could not. It turned out that it had pulled out Barney's dentures and couldn't understand why Betty's wouldn't come out. When Betty explained that some people lose their teeth as they age, the creature asked her to explain aging because it couldn't understand the concept. But perhaps the most important part of the story that Betty related to Dr. Simon was the star map, a drawing that she was able to make under hypnosis of a map shown to her by one of the creatures to show her where they had come from. This star map, a depiction of an unknown part of the heavens, would, years later, become an important piece of evidence that Betty would be able to provide to substantiate her story.

At first, when the star map became public in John G. Fuller's *Interrupted Journey* with an introduction by Dr. Simon,[21] about Betty's and Barney's two lost hours aboard a space ship, astronomers had not yet discovered the existence of the twin-star Zeta Reticuli system. But Ohio elementary schoolteacher and amateur astronomer Marjorie Fish was intrigued by the map that Betty drew. Betty had said under hypnosis that she had seen the map on the space ship and the aliens had described it as their home and indicated different "trade routes" to different sectors of that section of the galaxy. These were stunning revelations, if true, and Fish set out to find the unknown star system.

Fish used Betty's descriptions and the map in Fuller's book to examine many different stars and systems, comparing them to the 1969 *Gliese Star Catalog*, a catalog not available to Betty Hill in 1961 or in 1963. Fish hypothesized that the system Betty saw on the map in the space ship was the Zeta Reticuli system, a twin-star system, as the home system of the aliens. This conjecture was debated back and forth for almost a decade and, interestingly enough, became the basis for a 1974 article in *Astronomy* magazine and generated much discussion. Thus, supporters argue, Betty Hill could not have known about the existence of Zeta Reticuli on her own because the existence of the twin star was not published until 1969, and the discovery of planets in that star system also did not take place until years later.

Another interesting element that Betty related to Dr. Simon was a medical procedure in which the creatures examining her inserted a thin rod into her abdomen. Now called "amniocentesis," but unknown to Betty at the time, it was a procedure to sample fluid in her womb, extracting DNA for fertilization purposes. The entities had also extracted sperm from Barney, an examination, which Barney did not remember, but which still caused him some physical reactions and emotional anxiety. These examinations, the stories of which were related under hypnosis, indicated that perhaps the Hills were being used as test cases for some sort of alien/human cross fertilization program.

Dr. Simon was very skeptical about the nature of the Hills' encounter with an alien nonterrestrial species, hypothesizing that Betty's nightmares of having been abducted by aliens and taken to their spacecraft suggested the same story to Barney, who repeated it under hypnosis. However, in an early 1990s interview with late-night radio talk show host George Noory on his *Nighthawk* show out of St. Louis, Dr. Simon stated that he believed that Betty and Barney believed the story and that they were not fabricating anything. In that interview, George Noory remembers, Dr. Simon didn't say whether he believed the abduction story or not, only that he believed that they believed, and that was as far as he would go. In an October 1967 article in the journal *Psychiatric Opinion* essentially endorsing the value of hypnosis for lost memory recall, Dr. Simon said he saw the Hills' experience as a psychological aberration. Because there is no alien abduction category in the DSM, the basic manual for psychological symptoms, Simon found himself in a tough place where he didn't want to be. He believed from what he heard in the sessions that Betty and Barney actually lived out these stories, but investing his medical reputation in an alien presence was professional suicide. So he took the "psychological aberration route" and relied on Betty's nightmares for Barney's story without ever having to deal with the physical evidence of the dress or the skin warts around Barney's genitals corresponding to where Barney said he was examined.

According to Kathleen Marden, "At first, however, Simon was dumbfounded by the fact that the Hills' hypnotic recall matched in so much detail. He even muttered during one hypnosis session, 'It can't be.' He was always searching for a viable alternative explanation. The best one that he could find was the 'dream transference hypothesis.' But Barney stated under hypnosis that he knew very little of the details about Betty's dreams and that they were only dreams and couldn't possibly reflect reality. He simply didn't believe what he had heard."[22]

Moreover, Marden suggested in private emails, "Had Dr. Simon stated publicly that he believe that Hills had experienced an abduction by non-human beings aboard a flying saucer, it would most assuredly have been professional suicide." However, in his personal letters to Betty Hill, Marden said that Dr. Simon was positive and fully supportive of Betty.

The news of the event gradually leaked out when Dr. Simon's transcriber, so astounded by the Hills' stories, handed the transcript to a reporter for the *Boston Traveler*, John Lutrell, who had heard of the interviews and the Hills' regression sessions. He published the story in the *Traveler*, a story that was picked up by the wire services, and Betty and Barney suddenly became the center of a publicity firestorm. They did not seek publicity, trying only to figure out what might have happened to them and why, when their private medical sessions became public knowledge. They ultimately told their story to John Fuller, who wrote in *Interrupted Journey* about their adventure. The book was optioned for a television Movie of the Week by actor/producer James Earl Jones, who played Barney. Actress Estelle Parsons portrayed Betty. Also, interestingly, it was the publication of Fuller's book that prompted the first public coverage of the story on radio by none other than famed host Long John Nebel at WOR. Long John hosted Fuller after having read the book and told him, "This is a good tale here, it's got a little of everything, and you will do well with it."

Over the years other evidence has arisen to support Betty's story in addition to the *Gliese Star Catalog* and the technique known as amniocentesis. Betty's dress, for example, which she preserved in a closet after her encounter, was analyzed by Dr. Phyllis Buddinger, an organic chemist, and found to have protein remains in the pink powder residue on the spots where Betty said the creatures grabbed her dress. Dr. Buddinger said that although she couldn't identify anything otherworldly about the organic chemical compounds in the powder residue, the simple fact that the pink residue showed up in twenty-four hours after Betty's claim of abduction, the correspondence of the powder residue with the torn dress zipper indicated that that someone or something was tugging at the dress, and trace of organic compounds in those places was suggestive of presence of life form. In fact, because there was no pink residue on other parts of the dress, Dr. Buddinger theorized that the pink residue came from something external to Betty and not from Betty herself.[23]

Moreover, by looking at the dress, both the control sample and the stained sample, through a high-magnification microscope, Dr. Buddinger told Dr. Ted Acworth on the History Channel's *UFO Hunters* she was able to see that the fibers of the stained sample had been pulled apart. There was greater spacing between them, an indicator that the dress had been under strain, probably from being pulled. That would account for the distressed and spaced fibers, particularly in the stained sample. Dr. Buddinger said it wouldn't actually be necessary to find some unknown protein-based compound on the dress even if the dress was touched by extraterrestrials. It was clear that the dress was contaminated by something, but the contaminants were quite common. Not at all otherworldly. Simply put, the universe is probably made up of all the same stuff anyway so the fact that there was no "unknownium" on the dress doesn't discount Betty's story in the least.

There was additional evidence left on the Hills' car, shiny spots, the size of silver dollars, Betty said, marked the spots where the entities touched the car as they tried to keep the Hills from escaping. And when a physicist friend of the Hills suggested that they try to get magnetic readings off their car where the spots were, they ran a compass over it. Betty said that every time they brought a compass near the spots, "the compass would just go spinning and spinning around." Barney said that when he moved the compass away from the spots, the needle would just "flop down," but when brought to the spots, the needle would start to spin. This, researchers have said, is another indicator that something was present on that car that affected the normal response to a compass.

The story of the Hills' encounter, although fiercely debated among skeptics and UFO investigators, still stands as one of the most amazing revelations of human encounters with the unknown. At the same time, however, the national sensation caused by the Hills' story, the involvement of the air force, the publication in Blue Book, and the official transcripts of the therapy sessions would certainly have made their way to the new President Kennedy, who, toward the sudden end of his presidency, would ask the CIA for a full release of the government's UFO files to the Soviets.

The Final JFK UFO Memos

On November 12, 1963, less than two weeks before he was assassinated in Dallas, President Kennedy wrote a top secret memo, which has only been recently declassified under the Freedom of Information Act, in which he directed the Central Intelligence Agency to declassify their UFO information in their files that might affect national security.[24] Referring to prior oral discussions with the CIA, the president indicated that he had asked James Webb to develop a program with the Soviet Union for the purposes of "joint space and lunar exploration."

Kennedy asked that the intelligence agency review prior UFO cases to mark those cases which might have presented "high threats" to national security and to identify cases from bonafide CIA and USAF sources. He asked this for the purposes of distinguishing between the "knowns and the unknowns in the event that the Soviets try to mistake our extended cooperation as a cover for intelligence gathering of their defense and space programs."[25]

He went on to write the next blockbuster paragraph, ordering the CIA, once the information was "sorted out," to "arrange a program of data sharing with NASA where unknowns are a factor. This will help NASA mission directors in their defensive responsibilities."[26] And he asked for the full review and report to be completed by February 1, 1964. Ten days later he was dead.

Maybe it was bad enough that Kennedy was under the influence of debilitating doses of injected liquid methamphetamine; maybe it was even worse that he was having psychotic breaks in public with the White House press corps in hailing distance. Or maybe it was even more worse than worse that he was setting up assignations with women of questionable backgrounds in private apartments in Manhattan where he was completely unprotected and could have been kidnapped and compromised. Or was it that he was taking significant doses of LSD provided to him by Mary Pinchot Meyer, his longtime mistress and friend from his Harvard days, who was murdered a year later in Georgetown and who was getting LSD directly from former CIA asset Dr. Timothy Leary.[27] But this document calling for a disclosure to another agency and ultimately to the Soviet Union of the CIA's most sensitive files, most sensitive documents, in a memo in which—and this is the big one—the president of the United States actually, in print, discloses the existence of UFOs.

On top of everything else, it was this memo that set the wheels in motion for what would be nothing less than an inside-the-palace coup.

The Days After Dallas

There are a few ways to commit the perfect crime. One way, of course, is to commit a crime with such a degree of stealth that no one knows that a crime has been committed. As good as that sounds, there is an old adage that nothing disappears without a trace or the truth will out, which makes the perfect crime always vulnerable at some point to an unseen flaw. Then there's a better way in which one commits a crime, gets indicted and goes to trial, and is found not guilty. We've seen that before, haven't we? Then there's best way: committing the crime and getting someone else to shoulder the blame, then get killed, then the person who kills him dies, then the case is closed forever. And that's what happened when the CIA broached the plan to Vice President Lyndon Johnson.

The offer, something the vice president could not refuse, was simple. Because of the growing fallout from the scandal involving Bobby Baker, Johnson's aide in the Senate, and the very real prospects that Bobby Kennedy was pushing to drop LBJ from the 1964 ticket, at the least, Johnson would fade into oblivion. At worst, with the Bobby Baker scandal and Johnson's own involvement, he could face jail time. Thus, an unnamed source from the CIA whispered into Johnson's ear, he could either go to jail or to the White House. His choice.

But it was a devil's bargain. In order for Johnson to get to the Oval Office, he would have to set JFK up for an assassination and then, when the deed was done, arrange to have a patsy take the fall, then kill the patsy while setting up a very public investigation to pin the crime on the patsy. Easier said than done, but LBJ had an ally who hated the Kennedys, especially Bobby, even more than he did. And that was FBI Director J. Edgar Hoover. If the CIA arranged for the triggers to be pulled, LBJ arranged for the cover-up and Hoover signed on to it.

In an October 12, 1997, article in *Newsweek*,[28] based upon his reading of the secret Johnson recordings from the Oval Office which were released by former First Lady Lady Bird Johnson, author, scholar, and MSNBC presidential historian Michael Beschloss writes that J. Edgar Hoover told President Johnson a week after the assassination that there was more than one Lee Harvey Oswald. This was also corroborated by retired Army Lieutenant Colonel Philip J. Corso, who discovered the dual Lee Harvey Oswalds in records from the United States Passport Agency.[29] Another Oswald, this one not fitting the description of the Oswald that had been caught in Dallas, was surveilled when he was at the Cuban embassy in Mexico to receive a cash payment. On the tapes, Johnson indicates that he is setting up a central commission, to be headed by Earl Warren, to investigate the assassination so as to preclude multiple investigations from a number of sources. Johnson and Hoover discuss the tentative members of the commission, specifically former CIA director Allen Dulles, but Johnson asks Hoover not to reveal what he knows about the multiple Oswalds to the commission.

A moment here to deconstruct this conversation. The chief executive officer of the United States and the director of the FBI, the chief law enforcement officer for the federal government, enter into an agreement not to disclose vital information about the assassination of the president of the United States to the very commission being assembled to investigate that assassination. This, at the very least, is a conspiracy between like-minded individuals to withhold evidence from a law enforcement body. That conspiracy amounts to an obstruction of justice. But it gets worse, much worse. Engaging in a conspiracy to cover up evidence in a homicide, both the president of the United States and the director of the FBI are accessories after the fact in a capital crime in the state of Texas. By their own admission to each other, they're guilty. And brought before the right judge, both of them could have faced the death penalty. That's how bad this was.

Chapter 11

LBJ AND UFOS

The Johnson administration was heavily punctuated by a series of UFO events, not the least of which were the Lonnie Zamora sighting outside of Socorro, New Mexico, in 1964, the Rex Heflin case in 1965, the crash of an unknown object in Kecksburg, Pennsylvania, in 1965, the possible UFO connection to the Northeast power blackout in November 1965, the Hillsdale UFO flap in 1966, and the Condon Report in 1967. And there were also UFO incidents in Vietnam during the war.

The Lonnie Zamora Incident

The April, 24, 1964, Zamora case from Socorro, New Mexico, that took place early in LBJ's term after JFK was assassinated, is still a hotly debated case over fifty years after its occurrence. It was studied, and taken seriously, by the military, corroborated by multiple witnesses and some trace evidence, heavily investigated by J. Allen Hynek, who said it was one of the cases that convinced him that UFO reports deserved greater scientific scrutiny, and, also, investigated by Professor James McDonald, who included it in his July 29, 1968, statement to the House Committee on Science and Astronautics.[1]

The case involved the report of witness Lonnie Zamora, a police sergeant in Socorro, New Mexico, who was in pursuit of a speeding black Chevrolet late on a Friday afternoon outside of town when he saw something that caught his eye and, ultimately, changed his life. It was a flame in the sky that accompanied the sound of a roar, or at least that's what Zamora thought it was at first, as he explained to air force investigators from Blue Book. He believed some explosives had gone off in the distance and, breaking off his pursuit of the speeding Chevy, went off to investigate, driving along a gravel road in the direction of the sound.

As he approached the area and watched the flame in the distance rise and then disappear, he still believed he was watching an ignition of explosives from a dynamite shack he knew was on the other side of the incline. His patrol car had trouble climbing the steep rise, but he finally made it and saw a small object shining in the distance about 200 yards away, he told air force investigators. Nothing anomalous, he thought at first, believing the object to be a car that had crashed. He also noticed two figures by the object, one of whom looked at Zamora in surprise. As he got closer, he saw that the object seemed more like an oval or the letter "O," he reported. The two people he saw seemed smaller than normal adults, but he thought nothing of it, assuming they were simply that: small adults. He called in a "10-20" location report to his dispatch and advised that he would be investigating the sound and the flame, thinking that it was a car accident and he would render assistance. Then, Zamora told investigators from Blue Book, and as it was reported in Blue Book, once out of the car he heard a loud roar that started low and rose to a high frequency and saw a bright blue and orange flame from under the object that had begun to ascend straight into the air, stirring up a cloud of dust.

Zamora ran away from the object, using his car as a shield, but kept the object in his field of vision as it rose. He later told his interviewers that it was an oval-shaped device, shiny like aluminum, and had no portholes or windows and no noticeable access port or door. However, he noticed some form of red lettering that he could not make out and an "insignia" that he did not recognize. It was not a military insignia nor an American flag. He ran from the object, over the other side of the hill, turned around, and watched the object continue to rise until it took off out of sight. He radioed a report of the sighting to the county sheriff, who tried to locate the object through his headquarters window, but could not see it. Then Zamora was joined on the radio by a state police sergeant, Chavez, whom he asked to join him at his location. Then, while waiting for Chavez to arrive, Zamora walked to the spot from where the object had taken off and said to his interviewers that he noted an area where the brush was still burning. When Chavez arrived, he accompanied Zamora back down to the area where Zamora had spotted the object so Zamora could show him he burned residue, but Chavez found something else, tracks in the dirt as if they were made by landing skids. Zamora reported that the oval object was standing on a set of what looked like four metallic legs on the bottom of the object. The marks in the dirt from these legs were also seen by other officers on the scene after Zamora had called in his report. Officers also noted that the brush was still smoldering when they arrived.

Zamora was not the only person to have seen the strange-looking flying oval sitting atop a flame. Other people claimed to have seen the object, one of whom not only saw the object, but saw a police car approaching it. Presumably, that car was Zamora's.

A formal investigation began within a day after the sighting, during which both Sergeant Zamora and state police Officer Chavez were interviewed at length by the air force and the FBI. Both agencies visited the site and collected physical samples from the soil, including fused sand, sand that had been melted by the application of great heat. It was this fused sand evidence that made its way to Professor James McDonald at the University of Arizona years after the incident.

By the time McDonald learned about the sand from analyst Mary Mayes, who had been asked to analyze trace evidence from the craft's landing site, the Zamora case was already hotly disputed. The military was debunking the sighting completely while J. Allen Hynek was calling it a vitally important incident. But it was Mary Mayes who grabbed McDonald's interest because she said that the military had given her their assignment of testing the soil for radiation—which she could not find—but did find unidentifiable organic remains at the landing site, burned plants that showed exposure to an anomalous agent that had dried them out completely, and the fused sand.

As for Lonnie Zamora, he ultimately became a person of interest for ufologists around the world as well as for debunkers, who claimed he was perpetrating a hoax. Journalists also kept asking him for interviews. Zamora quit the police force, quit talking to ufologists and journalists, and became a gas station attendant, passing away in 2009. But during his years after the sighting debunkers like Donald Menzel from Harvard, whom UFO historian and nuclear physicist Stan Friedman said was a member of MJ-12 and debunking was his cover, said that Zamora didn't know what he was looking at or that some students were playing a joke on him and confused him into thinking that he was seeing something otherworldly.[2]

Actually, Zamora thought that he had seen a top secret test craft from nearby White Sands. CIA asset and UFO debunker and CIA embed Phil Klass said that either Zamora only saw ball lightning or that he himself was perpetrating a hoax to get more tourists to Socorro. Professor McDonald blew holes in both of these accusations, particularly the ball lightning theory. Project Blue Book deemed the sighting an "unknown" even though they dismissed any extraterrestrial origin for the object. The head of Blue Book, however, remarked that Zamora was an excellent witness, highly credible, who really did see something whose origin was identifiable. The investigation continues to this day.

Back in 1968, however, at the end of President Johnson's term, the president of New Mexico Tech in Socorro, Dr. Stirling A. Colgate, who died on December 1, 2013, wrote a letter to Nobel Prize winner Linus Pauling, in response to Pauling's inquiry about the Zamora sighting, in which letter Colgate said the Zamora incident was a hoax, nothing more than a prank perpetrated by students at New Mexico Tech. Dr. Pauling reportedly had an interest in UFO sighting reports and wanted to know what Colgate thought of the Zamora story. Colgate wrote Pauling that he had firsthand knowledge of the prank and that Zamora had been completely fooled by it.

There are problems with this story, however, not the least of which is that, assuming for the purpose of argument that everything Dr. Colgate wrote is absolutely true, there is no dispositive proof that what the students were hoaxing was what Zamora actually saw. What if they did indeed perpetrate a hoax, but Zamora saw a real object? Colgate's letter, although persuasive, is really a classic example of the *post hoc ergo propter hoc* fallacy. Just because some students were pulling a prank, possibly in the same area where Zamora was patrolling, doesn't mean that it's the cause of the Zamora sighting.

Another explanation might involve the testing of equipment for upcoming NASA missions because, at least to some UFO researchers, the object that Zamora reported seeing was far too conventional for some type of otherworldly device. The object was an ovoid capsule-shaped device supported by landing skids or legs and ascended via a type of rocket propulsion exhaust system. Why would extraterrestrials, presumably more advanced than us, rely upon a conventional rocket propulsion system, circa World War II vintage, on such a small device? Again, this doesn't provide a clear explanation for the Zamora sighting because if the air force knew that an aerospace company was testing a devise for NASA, wouldn't that have been the explanation in the Blue Book report? And wouldn't J. Allen Hynek have been apprised of this conventional explanation? Hence, the case remains a mystery.

The Zamora sighting in McDonald's file is one of the most detailed cases he followed. McDonald noted not only Zamora's specific descriptions of what he saw—the object's size, appearance, speed, and behavior—but the reports of marks in the sand left by the object as well as fused sand and scorched brush. Numerous witnesses to these traces were on scene within minutes of the sighting. After Zamora called in his report to his dispatcher, a fair number of law enforcement agents were at the scene within ten minutes of the radio call. They went to the actual landing site. These were law enforcement personnel, who could confirm the trace evidence Zamora reported, such as the landing marks in the dirt and that there were burn marks in the brush.

Because there have been so many different opinions about what Zamora saw and so much contradictory theory about the nature of the thing that he saw in the desert, on *UFO Hunters* Pat Uskert decided to get the story firsthand from the witness himself.[3] Lonnie Zamora, who had not spoken about the incident for years, had agreed to break his silence and talk to the camera. Allowing Zamora to recount his experience in person would reveal what the police officer actually saw and what he believed had happened. But could Zamora be believed?

According to Socorro police chief Lawrence Romero, "Lonnie is an upfront type of guy, an excellent police officer, and a pillar of his church."

Zamora had not talked to anyone about his sighting experience in over a decade, but he told the UFO Hunters that he wanted to tell his side of the story again and explain why he believed his sighting was real and not his mistaking conventional event for something unearthly. Zamora said that from where he first stood, he could see down on top of the object. "It was sort of an egg-shaped object about thirty to forty feet wide," Zamora remembered. "I didn't know what it was until I started moving towards it. Then I realized it was something I'd never seen before. I thought it was some kind of air force experiment, I thought at the first."[4]

The odd shape and structure of the object that Lonnie Zamora described was similar to NASA's lunar module. In fact, air force investigators theorized at first that what Zamora saw might have been a test of this same vehicle. But further analysis quickly proved this theory wrong. A flying, working model of the lunar module was not ready until 1965, a year after the Zamora sighting. However, what had landed in the Socorro desert had left physical evidence of its presence, indentations from the object's landing skids in the desert sand.

"All around here," Zamora said, pointing to the area where the object had stood almost fifty-five years before. "In fact, there is one right here. And there's another one right over there where the rock is. And another one there and one more right here." Zamora outlined the area where the object had sat. "There were four," he said.

Interestingly, the indentations were not the only traces of physical evidence left by the object. According to Zamora, bushes were burning and rocks were smoldering after the object he said he saw had lifted off from the site. Zamora pointed to four clumps of bush. "But there was one that was burning so hot," Zamora said, standing over the burn spot, "that you couldn't get near it."

Zamora said that the burning scrub on the desert floor landing spot was real evidence that something had been there, but the evidence was quickly confiscated by air force investigators, who arrived soon after Lonnie had reached the site and then made his report to his supervisors. However, James McDonald revealed in his review of the case that it actually wasn't the air force that had collected the debris and asked that it be taken away. It was the FBI. And adding to the mystery of this incident, the FBI, according to McDonald's files, didn't want anyone to know they were working on the case. The FBI agent, D. Arthur Byrnes, had accompanied air force captain Richard T. Holder to the site and, according to Captain Holder, requested, as indicated in McDonald's notes, that "no mention be made of the FBI's interest in the case."

The reason for this request has never been revealed, but one can only assume, given J. Edgar Hoover's almost obsessive micromanagement of his FBI and his close friendship with President Lyndon Johnson, that whatever Hoover was told by Agent Byrnes was reported right up the ladder to President Johnson because UFOs by this time, especially after the Kennedy memo to the CIA and national security agencies, were serious business.

FBI agent Byrnes and air force captain Holder were excited about the event and the trace evidence at the scene. When Zamora tried to ask questions of the investigators, he recalled, they refused to answer and, according to Zamora, "they wouldn't let me ask them any questions and told me to keep it quiet for awhile." But James McDonald wanted to break the silence about the incident and doggedly tracked down specialists who had been brought in by the government to investigate Zamora's sighting.

McDonald's files indicated that Mary G. Mayes from the University of New Mexico was enlisted to analyze plant material from the site. In her interview with McDonald, Mayes described the location and reported seeing a twenty-five to thirty-foot patch of fused sand at the site where Zamora said the object was sitting. According to Mayes, a small area of desert had turned to glass. This was additional trace evidence that would require a test to determine how that sand was fused into glass and whether the fused sand might shed some light on Zamora's story. The question would be, how much energy would it have taken to fuse the sand from the site? Could the sand have been fused naturally? Might the prank that New Mexico Tech president Dr. Colgate described to Dr. Linus Pauling have generated enough energy to fuse the sand into glass?

Patric Morrisey, a glass blower who agreed to work with the UFO Hunters, tested sand from the site to provide the answers. Did something generating heat high enough to fuse the glass take off from that spot or did pranksters from New Mexico Tech fuse the glass themselves to make it look like something had blasted off from there? A demonstration of the effects of the level of heat required to fuse the sand would test out the two theories.

Morrisey took the sand samples and set them on a fired clay base that could withstand the intense heat of a glory hole, the furnace where glass is melted so that the artist can form it into shapes. While the sand was heating, Morrisey explained that if hoaxers brought a blowtorch to the site to fuse the glass and the blowtorch put out a strong flame, it still would not fuse any sand into glass at all because the heat is not intense enough. It would take an actual oven to fuse sand into glass. Or it would take a nuclear blast.

Morrisey explained that the bonding of the sand in an oven didn't mean that the sand was changing into glass because it would take a temperature of at least 2100 degrees Fahrenheit applied for at least a half an hour to fuse the sand into glass. Moreover, the glass trace evidence from fused sand, had the pranksters brought glass to the prank site in order to hoax someone like Zamora, would have had to have been the same composition as the surrounding sand at the site. More than likely, had hoaxers provided the fused said, it would have had a different composition. Hence, if the fused sand at the site was of the same composition as the surrounding sand, it is likely that trace evidence was not the result of the pranksters. Thus, Morrisey said, the hoaxer scenario would be very hard to believe. Therefore, the Stirling Colgate student practical joke explanation fails on its face while an examination of the physical trace evidence lends a great credibility to Zamora's experience.

The Rex Heflin Photographs

The Rex Heflin photographs have over the years become a kind of Holy Grail for UFO researchers not just because of their clarity but because of the series of events that followed the disclosure of these photographs, disclosures that went to the very heart of the United States military hierarchy and almost surely, because of national security concerns, made their way to a presidential briefing. We know, not just because of the Kennedy memo, but because of records from both the Ford and Carter administrations, now public, that dramatic UFO incidents were briefed to the president.

The events began on August 3, 1965, in Santa Ana, California, at approximately 12:30 p.m. when traffic investigator for the county road department Rex Heflin was about to take a photo of a tree branch that was obscuring a road sign.[5] But out of the corner of his eye, he witnessed something fascinating. A saucer-shaped craft with a round structure on top was moving across the horizon at 150 feet above the ground. He was transfixed and told his co-workers that he pointed his Polaroid 101 instant camera at the object and snapped off a series of three shots of the object.

As the object traveled left to right across the road, the first photo, taken through Heflin's front windshield, shows a circular object, oddly looking like a flying disk-shaped hubcap, that appeared to be off in the distance flying above the ground at the edge of the roadway. The second photo shows the same disk taken through the passenger-side window of Heflin's truck. In the third photo, also taken through the passenger-side window, the object looks smaller as if it were flying away from the camera. Within two minutes of the initial sighting, the object disappears, leaving behind a peculiar donut-shaped smoke ring floating in the sky.

The other interesting incident that occurred during Heflin's sighting in Santa Ana involved the radio in his van, which suddenly was blanked out by electrical interference when the object appeared and passed in front of the vehicle. After the object disappeared from view and only left the floating smoke ring, the van's radio came back to life. Heflin's boss at the Orange County Road Department offered no explanation for the performance of Heflin's radio, which checked out as perfectly functional after the incident. The only possible explanation for the interference would have been that Heflin had keyed the mike on his two-way, thus sending out a carrier signal that would have blocked any incoming transmissions. But Heflin said he did not do that. A similar type of electrical interference with an electrical device would occur in September of that same year when Heflin was visited by two men who said they were from NORAD and were investigating the incident.

The photos Heflin had taken were heavily investigated by all branches of the military, and thus, at some point, their existence would have been made known to the office in the Pentagon where such anomalous investigations are conducted. The photographs also came to the attention of Professor James McDonald, who wasn't the only person fascinated by the Rex Heflin case.

The air force, too, investigated it as well as Project Blue Book, and the infamous Condon Committee investigated it. All of those groups dismissed the case as a hoax. For the record, though the Condon Committee's 1967 findings, mainly the executive overview of Dean Edward Condon at the University of Colorado, said that UFOs, which were more illusion than real, an investigation into them was unlikely to yield any serious scientific results, and, in any event, posed no threat to national security and, thus, were not worthy of investigation by the air force. That moved the air force to close down Project Blue Book in 1970, during Nixon's term in office. But none of the Blue Book conclusions satisfied James McDonald, who was convinced that not only were Dean Condon's conclusions deeply flawed and showed a blatant disregard for the real evidence presented by some UFO cases, but that more scientific study was needed about the Heflin case in particular.

There is another weird aspect concerning the self-proclaimed NORAD representatives who visited Heflin to take his original photos. Prior to the NORAD representatives showing up at his door, in September 1965, Heflin received a phone call from someone identifying himself as a NORAD officer, asking him for a meeting so Heflin could give NORAD the photos for analysis. The person on the phone strongly advised Heflin not to talk about his sighting or the photos with anyone. When the two individuals, dressed in civilian clothes, showed up at Heflin's house a few days later, Heflin said that only one of them came to his door. The other stayed in the car, but, strangely, when Heflin looked at the car, he saw a purplish glow coming from inside the car, a glow he could not identify, but which raised no obvious suspicions at the time. As the one representative stood in his doorway, Heflin also noticed that his high-fidelity radio receiver inside his house suddenly cut out, blanketed by some type of interference. At that moment, Heflin did not immediately connect the interference on his hi-fi with the interference he experienced on his van's radio when the object in his photos flew across the road.

After Heflin claimed his original photos were not returned from NORAD, the air force made public its review of the photos, officially labeling them a "photographic hoax." With only copies of the originals to analyze, some debunkers claimed to see a line just above the UFO indicting some type of wire or connection to the object. Others disagree, saying these are alterations or defects seen only in the copies. Without the original photos, there was no way to be sure. Then, suddenly, other copies of the photos begin to spread through the UFO community, copies, James McDonald's biographer told the UFO Hunters, that had been altered.[6]

Thus, if one were paranoid, one could say that there was some sort of a disinformation program going on with various copies of the photos that kept turning up. Some conspiracy theorists believe that the originals of the photos were taken by the men announcing themselves as NORAD representatives for precisely this reason, sequestering the originals and distributing copies that had been altered to debunk the originals and McDonald himself. But then, almost twenty years later, something miraculous happened to McDonald, when one day his phone rang. At the other end there was a woman's voice asking, "Have you checked your mailbox lately?" Heflin was puzzled, but he went out to his mailbox where he saw an envelope tucked inside. When he opened the unmarked envelope there were the three original photos that the so-called NORAD men had borrowed twenty years before.

Ann Druffel is the closest source to the original photographs taken by Heflin, and has said that Rex Heflin died in 2005 and never wavered in his story that he had photographed an unnatural object, a real object, but very unusual. But the question remained, did Heflin hoax the photos or were these images real, was this a real object he captured on his Polaroid camera? A digital analysis of the closest copies of the originals was conducted to determine whether the line that debunkers claimed was holding up the object was real or it was simply an anomaly on a copy of the original photo. Another theory that required testing out was the argument, set forth by debunkers, that the 1950s-era Polaroid 101 camera Heflin was using could not possibly have snapped off the first three photos in the twenty seconds that Heflin said he took. The entire incident happened too quickly, debunkers have argued. The Polaroid, which registers the image inside a cartridge within the camera, was too slow to catch that entire sequence as it was happening because its mechanism was still spitting out the developing photos while adjusting the lens aperture to focus on taking new ones. Thus, the photos had to be staged and hoaxed. A test was made to ascertain speed of the camera with Ted McClelland, a technical expert from Polaroid, who used the exact same model Polaroid that Rex Helfin used in his everyday work for the Orange County Road Department and that he used to take the series of photographs of the object that looked like a flying saucer.

"As in most Polaroids," McClelland explained, "the photo package goes into the back of the camera. When the back film door is closed the camera is locked tight, admitting no light except through the shutter to expose the film. When the operator clicks the shutter, he or she has to slide the frame out of the back of the camera slot immediately so that the chemicals can process the exposed film frame. The frame includes the exposed photograph so that the actual developing takes place outside the camera, not inside."[7]

Ted McClelland conducted the first acid test of the camera's ability to snap off three shots in twenty seconds or less. He snapped the shutter while a time clock was running. McClelland was able to snap four photographs in seventeen seconds, beating the twenty-second three-photo requirement.[8]

This meant that an operator, Rex Heflin, could have caught all three images of the object crossing the road in front of his truck. Therefore, debunkers were wrong about the speed at which Heflin claimed to have taken his photos of the object. This debunking claim had dropped off the list.

The next controversy for debunkers was the sharp focus of the object on the Heflin photographs. Was it too focused, too sharp? Did the sharpness of the object in the photos suggest a double exposure, thus a deliberate hoax on Heflin's part? A double exposure on a Polaroid 101 means that two photos are taken on the same frame of film so that one image is imposed upon another. If this is what Heflin did, he would have had to take a first photo of the sky through his windshield and not remove the frame. Then he could have thrown a hubcap into the air and taken that photo on the same frame, which he kept in the camera. Thus, the image of the hubcap would appear in the frame of the sky, looking as if the hubcap were flying through sky so that it resembled a flying saucer. Debunkers have said that this is exactly what happened because the big clue was that every part of the through-the-windshield was in focus, the object, the background, and the frame of the interior truck windshield. How could this be explained?

Depending upon the film stock, the distance from the lens, and the distance to which the lens is focused, objects in different parts of the film, foreground and background, can appear out of focus. In motion pictures and in many action photos, especially in sports, the photographer deliberately focuses on a specific object as his or her subject so as to render the background out of focus. How many times in a movie have we seen focus on a background object in a frame and then the operator changes the focus to bring into focus an object in the foreground, thereby deliberately taking the background out of focus? Heflin's ability to capture a crisp foreground and background led many to believe he had set the shot up as a double exposure.

"Setting the camera lens to infinity and using 3000 ASA (film speed) speed film, you would be in focus from three feet to infinity," Ted McClelland explained.[9] In this way, Heflin's camera could have kept all the disparate objects in the film in focus at the same time as well as the center line on the road going off into infinity. Therefore, the camera was capable of keeping everything in focus. This possibility debunks the debunker argument that because everything in the photo was in sharp or crisp focus, it meant that the photos were double exposures. The photos didn't have to be double exposures to keep all the objects in the frame as sharp and crisp images.

Next debunker argument up concerns the fourth photo of the smoke ring, which seems to show a smoke ring that lingers after the object, according to Heflin, flew off. Even Professor McDonald had a problem with the fourth photo because the first three photos of the image, taken from inside the truck, show a clear sky in the background. But the fourth image of the smoke ring, taken from outside the truck, shows a cloudy sky in the background. What could have accounted for the difference in the background? McDonald thought at first that these were hoaxed photos because of the difference between the fourth photo and the first three. He couldn't bring himself initially to accept that the photos had actually captured a flying saucer in flight even though Heflin said that he took the photos in succession and that the object in them was real and had left a smoke ring when it flew away.

Ted McClelland demonstrated that the missing part of the explanation probably resulted from the automatic exposure setting on the Polaroid 101, which takes into account the ambient background light and adjusts the image based upon the light. Thus, because the first three photos were taken from inside the van, the camera might have been tricked by the darkness inside the van, where the automatic exposure reacted to that darkness and overcompensated by opening the lens wider, thus allowing more light to flood the frame from outside the van. However, "in the last photo, there was nothing to interfere with the automatic exposure and the camera is picking up exactly what it sees," Ted said.[10] Because, now outside the van, Heflin's camera picked only the outside light, it took a true image of the cloudy sky that it could not do because it overcompensated for the lack of light inside the van. That would explain the difference in the background between the fourth and the prior three photos. Hence, according to Polaroid expert McClelland, the mystery is explained and the difference in background is accounted for.

The fourth debunker argument is the harshest of all because it asserts that the object itself is a fake, a hubcap or something similar, and not a flying saucer. Project Blue Book, for example, labeled the photos as a hoax, calculating that the object in the photos was only nine inches in diameter, twelve feet off the ground, and about fifteen to twenty feet in the distance from the van. McClelland disagreed with Blue Book, saying, "There is no hoax that I saw. Holding up the negatives and magnifying them as best as I could, I saw no signs of a hoax."[11] There were no strings, no piano wire, no external support for the object over the ground, nothing that spoke of a hoax to McClelland's expert eye. However, the strongest evidence that the photos were not hoaxed lies in the calculations made by McDonald himself, calculations that a digital photo analyst could make and display visually. The photo analyst who worked on these photos was Terrence Masson.

Heflin claimed the object he photographed was roughly twenty feet in diameter and about an eighth of a mile from his truck or 700 feet from the lens of his camera. Project Blue Book said the photos were hoaxed, the object was small, a hubcap, no more than nine inches in diameter, maybe only twelve feet off the ground, and close to the camera, perhaps fifteen to twenty feet away from the lens. Using Heflin's and Blue Book's numbers, Masson on his photo analysis program constructed a three-dimensional landscape to compare Blue Book's conclusions with Heflin's observation and test his claims. Digitally, he placed a twenty-foot-diameter saucer 700 feet in the distance, matching Heflin's description of the incident. If Heflin had hoaxed the photos regarding the size and distance of the object, as Blue Book and many debunkers have said, there would be almost no chance of the three-dimensional model matching the photos because Heflin's numbers wouldn't comport with the image.

"If he was telling the truth," Masson said, "and he was making an accurate eyeball estimation, that object in our simulation the object has to line up with our twenty-foot-diameter object, 150 feet off the ground."[12] Terrence punched in the parameters and began sliding the actual photo image of the object to meet the parameters of the 3-D model. Then he cross-faded the model to the Heflin photograph, and guess what, the image comported almost exactly with the model. The debunkers were, in Terrence's words, "not even close." The image and the 3-D model based on Heflin's measurements lined up perfectly, meaning that Heflin was right and Blue Book and the debunkers were wrong in their estimates. The photos were not hoaxed. Worse, it was the government that was the hoaxer.

"This is just very simple trigonometry," Terrence said. "You just put in the space that's very accurate at the size of the object and the height of the object that Heflin said it was and his estimate and the actual photographs match very, very precisely."[13]

The conclusion has to be, therefore, based on the analysis of the photo, that Heflin had to have been observing something not under his control. And the other tests on the camera indicated that the camera was capable of not only snapping off the number of photos during the time sequence Heflin said had transpired, but was capable of keeping all elements of the image in focus if the lens had been set to a distance of infinity. And because of the automatic light exposure control in the Polaroid 101, the ambient light entering the lens while Heflin was photographing the object from inside the truck cab overexposed the sky background, blocking out the clouds. Once Heflin stood outside the cab to photograph the smoke ring, the camera adjusted to the background light and captured the images of the clouds. Thus, the debunker arguments were incorrect and the Blue Book evaluation of the photos was not only mathematically incorrect, the Blue Book analysts actually hoaxed their debunking. Is this what was truly reported to President Johnson?

The Great Northeast Power Blackout

On November 9, 1965, shortly after five in the afternoon, as the sky was growing dark, lights flickered in buildings in Manhattan, came back on, and then shut off. Folks looked around at the tall buildings whose lights were out as well. The subways had stopped running, traffic lights had shut off, and restaurants began putting out their food for free. Manhattanites wandered the streets, especially in downtown, looking for answers but worrying that this was, indeed, the end of the world. In New York's outer boroughs, it was much the same story as it was for people in Manhattan trying to get home from work.

For folks who had transistor radios and could pick up stations from outside New York, it was much the same story. And Washington, DC, too, was thrown into darkness. The outage was so extensive that President Johnson suspected foul play. In fact, he assembled his national security experts to evaluate whether this was a Soviet-based event, an act of sabotage that preceded a ballistic missile attack, or, maybe, the work of a UFO. Tensions were running high. LBJ, also suspecting that a domestic crime had occurred, called in J. Edgar Hoover at the FBI to connect as best he could with his

Canadian sources to determine if there had been any sabotage along the northeast power grid.

Although in the immediate aftermath of the blackout, there were fears at the highest levels of the Johnson administration, including the president, that the Soviets had managed to compromise our grid, engineers ultimately found the real cause: a relay in the Ontario, Canada, grid became overloaded and failed, allowing the power surge to overload other circuits along the northeast grid and as each grid failed, the entire northeastern part of the US and Canada were thrown into a complete power loss.

It wasn't the Soviets or a UFO, but for a brief period of time, our readiness for war was under scrutiny even if the defense chiefs were not sure whether the Soviets initiated the blackout or whether they were ready to take advantage of it.

The Crash at Kecksburg

The event began on December 9, 1965, at about 4:45 p.m. when a glowing fireball swept over the midwestern United States, crossed the Ohio border, and appeared over western Pennsylvania. Surely this object was picked up on radar, because we know its course, and surely even a perfunctory warning would have reached the national security team at the White House, if only because this incident report came just a month after the great northeast blackout.

But observers looking at the spectacle across three states said that it didn't act like a descending out-of-control fireball because it was moving too slowly. But then, according to witnesses, the object seemed to lose control of its landing momentarily and then fell to earth in a ditch outside of rural Kecksburg. What happened then had an strange similarity to what happened in Roswell less than twenty years earlier.

One of the best experts on this case is Stan Gordon, the primary investigator on this case, who became involved all the way back in the 1970s. Stan began with the collective witnesses' observations that the object appeared overhead as if it were navigating and coming down in a controlled landing. And as it fell into a thicket of woods, there was a bluish smoke rising out of it but dissipating into the cloud very quickly. Stan said, as did John Ventre from Pennsylvania MUFON (Mutual UFO Network), that this object was likely reported in the skies over Ontario, Michigan, and Ohio before it crossed into western Pennsylvania. As it was falling, smoking and brightly illuminated, Stan said on the *UFO Hunters* episode on "Alien Crashes" from season 2, "The police department, the fire department, and radio station were all being bombarded with calls. And as it proceeded into Westmoreland County, local county first responders were overwhelmed with calls coming in."[14]

The entire town of Kecksburg was consumed by the incident, before and after the landing and in the weeks and months after it. The images of the event have stayed with witnesses for decades, according to Billy Buelbush, who was one of the first witnesses to the object, watching it fly over and crash. "Heard this sizzle noise," he said. "Then I seen this red fireball coming. I watched it come over and heads for the mountain. Then I saw it turn, a perfect turn, and come back. It came just like it was looking for someplace to land." Pat Uskert asked if the object were under intelligent control and Buelbush said that "it had to be."[15]

Buelbush said that he drove to where he believed the object, which he thought at first was a downed aircraft, had come to rest. He said that he pulled his flashlight from his car and walked over to the impact spot which was in the woods. "I could see it down there," he said, but wondered how he could extract the occupants because "there were no windows, no doors, no rivets." Standing approximately eight feet away from the object, Buelbush said it smelled like sulfur, smelled like it was made of rotten eggs. The general consensus among witnesses, over fifty years ago in Kecksburg, was that they were impressed by the description of the object's turning, looking for a place to come down, which meant that it was under intelligent control and not just tumbling out of the sky. At the site, immediately after he discovered the impact site, Buelbush was joined by units from the Kecksburg Fire Department, responding to reports of a burning crashed object in the woods.

Volunteer firefighter Jim Romansky reached the object first, surveying the damage he believed he would find. "I was shocked," he said. "Because I was prepared for a smashed up airplane, and here was this humungous piece of metal buried into this drainage ditch. Buried into brush because it came flying down from the trees."[16] It was acorn-shaped, Romansky said, and had a ring of strange markings that circled the entire object. This object, however, displayed none of the traditional features of UFOs. It looked like an artillery shell or a large bell, unlike flying triangles, discs, or cylinders.

If anything, it resembled the shape of the object that flew into Rendlesham Forest outside of RAF Bentwaters fifteen years later. The strange graphics on the side of both the Rendlesham and Kecksburg objects made for a chilling similarity, which became even stranger when one considers the strange graphics that the late Jesse Marcel Jr. saw on the debris that his father brought back from the Roswell debris field in 1947.

"I'd never seen anything like this in my life," Romansky said. "My impression after I looked at this thing, after I walked around it, after I thought about some more as I stared at it was that this thing wasn't from here."[17]

Within hours after the crash a contingent of military personnel arrived, including both army and air force, fully equipped with retrieval and transportation apparatus. They cordoned off the area around the drainage ditch where the object lay, still smoking, still glowing with a purple hue, where they prevented the public from viewing what was in the ditch and pushing away those already assembled at the site and keeping others outside the area. Inside the newly restricted zone, the military began searching. Bob Gatty, a reporter from the *Greensburg Tribune*, a local paper from the next town over, was dispatched to the scene by his editor to cover the event. He said that he remembered very clearly the heavy and intimidating military presence. "There were at least a dozen military personnel in addition to Pennsylvania State Police." He said, "'I need to go down there. And they said, 'there's nothing down there.' I said, 'if there's nothing down there then why can't I go in there?'" He said that his impression was that the military and not the state police were in charge of the scene. "I had never experienced a situation in which the military had come into a civilian situation and taken away control of an accident site. Why would the army be there?"[18]

Witness P. David Newhouse said that as they walked up to the site they were stopped by the military. "I heard the bolt snap on the rifle. We stopped and talked to the soldier for a second, and he told us to get out and get back to where we came from." Witness Robert Bittner said that a soldier looked him as he approached the site, pointed his finger at him and said, "Nobody goes down here." And Bittner turned around and left the immediate area. Witness Billy Buelbush also claimed to have seem officials from NASA wearing white hazmat disposal uniforms at the retrieval site. Later that evening, however, according to Stan Gordon, witnesses reported seeing a large flatbed truck enter the area. In an hour, the truck left from the site in an obvious hurry carrying something under a tarp.[19] And Stan Gordon said that years after the incident itself, he conducted interviews independently and separately with Billy Buelbush and Jim Romansky, both of whom led him to the site where they remembered seeing the object. It was the exact same location where he believed the object had landed. And he said that a search for any physical evidence might also resolve an issue that debunkers had been proposing for years, that the Kecksburg object was really part of a Soviet space probe, the Cosmos 96 Venus probe, that fell back to earth. If the object was Cosmos 96, its tiny nuclear reactor might have left traces of radiation whose signature might identify it, although other evidence had eliminated the Cosmos 96 as the crashed object. The National Space Data Center, for example, said that according to air force tracking data, Cosmos 96 entered the atmosphere and was destroyed earlier than the first sighting of the Kecksburg object. Hence, it could not have been the Cosmos 96 probe that fell out of the sky into the Kecksburg site.

Over the years rumors have circulated, some alleging that the Kecksburg crash was just a false story while others said that NASA did visit the site right after the crash, accompanied by a black ops military team to scrub the site and keep witnesses from talking. Eventually, John Podesta, a member of President Clinton's administration, one of the managers of President Obama's transition team, and a Hillary Clinton 2016 campaign senior official, came forward to demand the government, and particularly NASA and the military, reveal the details of their retrieval of the Kecksburg object and their findings concerning the nature, origin, and purpose, if any, of the object. Journalist and author Leslie Kean filed under FOIA for information about the Kecksburg crash and what NASA discovered.[20] But her organization, the Coalition for Freedom of Information (CFI), said that NASA was still hiding important facts about the crash. NASA had first stonewalled the CFI's FOIA request, but, after the CFI filed a lawsuit to compel the government to release whatever Kecksburg information it had, NASA released some information, but the information raised more questions than it answered, specifically information relating to the filing of Kecksburg information in a file called "Moondust," a government program, referenced in State Department documents, under whose auspice unidentifiable objects from space that fell to Earth were cataloged.

What Kean believes she found out dispositively eliminated the Soviet Cosmos 96 based on the US Space Command and Soviet Space Agency records. Therefore, if not Cosmos 96 or a meteor, what was it? The NASA documents offer no affirmative clue, but, according to some UFO historians, the object that crashed in Kecksburg might have had its origins in Germany and then in Poland in the 1940s as a Nazi military experiment that succeeded all too well, especially inasmuch as the director of the 1940s Nazi Bell project, Die Wunderwaffe, was an electrical engineer, and SS Colonel Dr. Kurt Debus, who was patriated to the United States under Operation Paperclip and became a director of NASA, was in charge of the retrieval effort at Kecksburg. A coincidence?

The Federal Government's Men in Black at the Kecksburg Crash Site

In Kecksburg, the most chilling story about information suppression involved local radio newscaster and reporter John Murphy from WHJB, who interviewed one of the first witnesses to the glowing blue descending object, Francis Kailp. Based on his first phone call with Kailp, Murphy called the state police, who arrived at the crash site to investigate. After arriving at the site, however, the state police unit informed Murphy that they were contacting the military, an explanation for the sudden appearance of a military security and transport team leading anyone to ask why a transport team. Then the state police told Murphy that they could find nothing on the ground and there was nothing to report. Another question, if there was nothing at the site—a statement that eyewitness testimony belied—why call in the military? It was a question that Murphy wanted to answer, especially after he heard a trooper on the phone talking to someone in an official capacity about the police discovery of a bluish purple pulsating object in the ditch. Murphy made a trip to the site, tagging along with a police unit, when, at the site, he discovered it was cordoned off and that he was stopped from going any further.

Convinced that something was being hidden by the authorities, Murphy continued his interviews with witnesses, recording them on audio tape and taking photographs, until he was confronted in his radio broadcast booth by two men who had identified themselves to radio station personnel as government agents. One witness who saw them talking to Murphy in the broadcast studio said she believed Murphy was being intimidated because she could see him shrink and withdraw from the two men standing over him. Murphy's plans for his documentary were quickly abandoned after that confrontation, replaced with a version that made no mention of the object that had fallen. People also noticed that Murphy was a changed man after the confrontation and the broadcast of his documentary, seemingly withdrawn and prickly to the point of being incommunicative. Four years later, in 1969, Murphy was killed in a hit and run accident. That crime was never solved. Was Murphy a victim of government men in black who had frightened him into altering his documentary and then, because he knew too much, eliminated him and the threat of disclosure of secrets they were assigned to protect?

What might President Johnson have thought about all of this, especially in light of what happened the following year when there was a UFO flap in Hillsdale, Michigan, reports of which made the national news and encouraged then Congressional House Minority Leader Gerald Ford to challenge the Johnson administration by demanding that Mendel Rivers of the House Armed Services Committee investigate the entire UFO question, with particular emphasis upon the Hillsdale sightings of UFOs and actual alien beings?

The Hillsdale, Michigan, UFO Flap, 1966

In March 1966, residents in Hillsdale in southeastern Michigan were multiple witnesses of UFO sightings in the nearby hills. Students at Hillsdale College also reported seeing a disc-shaped object hovering above the hilltops and treetops and then descending to the ground, their landing spots behind the hills. According to UFO reporter and blogger Jim Cohen, for two and a half hours in the early morning on March 14, 1966, "Washtenaw County sheriffs and police in neighboring jurisdictions reported disc-shaped objects moving at fantastic speeds and making sharp turns, diving and climbing, and hovering. At one point, four UFOs in straight-line formation were observed. Selfridge AFB confirmed tracking UFOs over Lake Erie at 4:56 a.m."[21]

According to the official log of Complaint No. 00967 signed by Cpl. Broderick and Deputy Patterson of the Washtenaw County Sheriff's Department:

3:50 a.m. Received calls from Deputies Bushroe and Foster, car 19, stating that they saw some suspicious objects in the sky, disc, star-like colors, red and green, moving very fast, making sharp turns, having left to right movements, going in a Northwest direction.

4:04 a.m. Livingston County [sheriff's department] called and stated that they also saw the objects, and were sending car to the location.

4:05 a.m. Ypsilanti Police Dept. also called stated that the object was seen at the location of US-12 and I-94 [intersection of a US and an interstate highway].

4:10 a.m. Monroe County [sheriff's department] called and stated that they also saw the objects.

4:20 a.m. Car 19 stated that they just saw four more in the same location moving at a high rate of speed.

4:30 a.m. Colonel Miller [county civil defense director] was called; he stated just to keep an eye on the objects that he did not know what to do, and also check with Willow Run Airport.

4:54 a.m. Car 19 called and stated that two more were spotted coming from the Southeast, over Monroe County. Also that they were side by side.

4:56 a.m. Monroe County [sheriff's department] stated that they just spotted the object, and also that they are having calls from citizens. Called Selfridge Air Base and they stated that they also had some objects [presumably on radar] over Lake Erie and were unable to get any ID from the objects. The Air Base called Detroit Operations and were to call this Dept. back as to the disposition.

5:30 a.m. Dep. Patterson and I [Cpl. Broderick] looked out of the office and saw a bright light that appeared to be over the Ypsilanti area. It looked like a star but was moving from North to East.

6:15 a.m. As of this time we have had no confirmation from the Air Base.

"Washtenaw County deputies B. Bushroe and J. Foster formally stated: 'This is the strangest thing that [we] have ever witnessed. We would have not believed this story if we hadn't seen it with our own eyes. These objects could move at fantastic speeds, and make very sharp turns, dive and climb, and hover with great maneuverability. We have no idea what these objects were, or where they could have come from. At 4:20

a.m. there were four of these objects flying in a line formation, in a north westerly direction, at 5:30 these objects went out of view, and were not seen again.'" [22]

Over a hundred witnesses reported seeing the same shaped objects making impossible hairpin turns, hovering in midair, and sweeping off at speeds greater than any airplane could have made. But as the days wore on and the sightings continued, things became even stranger when witnesses reported that at least one of the objects actually landed in a swamp and other witnesses described a physical circular object sporting colored lights hovering over a house. Then a police officer reported that at least one of the objects buzzed his police car. And as the unidentified craft came close to the ground, the "Washtenaw County Sheriff Douglas Harvey ordered all available deputies to the scene. Six patrol cars, two men in each, and three detectives surrounded the area. They later chased a flying object along Island Lake Road without catching it."[23]

The number and official status of the witnesses—police, sheriff's officers, local community leaders—made this sighting all the more credible, especially the details regarding the circular nature of the objects, their ability to hover in place, and the swirling lights that were clearly not aircraft navigational lights or beacons. The intensity of the sighting was so great that the air force sent in its UFO expert, J. Allen Hynek, to investigate. Although Hynek would eventually come to agree that UFOs were a real phenomenon that had to be investigated, in 1966 he simply dismissed the sightings as "swamp gas," and that's how UFO sightings would be officially characterized for the ensuing decades.

Finally, on March 28, 1966, Gerald Ford, who would become vice president under Richard Nixon after Spiro Agnew resigned and then president after Nixon resigned, wrote to Chairman Mendel Rivers of the House Armed Services Committee and Chairman George Miller of the House Science and Astronautics Committee to call for an investigation of the Hillsdale sightings, a request that must have surely reached President Lyndon Johnson, who kept a very close eye on the goings-on in the legislature.

Minority Leader Ford wrote:

"No doubt you have noted the recent flurry of newspaper stories about unidentified flying objects (UFO's). I have taken special interest in these accounts because many of the latest reported sightings have been in my home state of Michigan.

"The Air Force sent a astrophysicist, Dr. Allen Hynek of Northwestern University, to Michigan to investigate the various reports; and he dismissed all of. them as the product of college-student pranks or swamp gas or an impression created by the rising crescent moon and the planet Venus. I do not agree that all of these reports can be or should be so easily explained away.

"Because I think there may be substance to some of these reports and because I believe the American people are entitled to a more thorough explanation than has been given them by the Air Force to date, I am proposing that either the Science and Astronautics Committee or the Armed Services Committee of the House schedule hearings on the subject of UFO's and invite testimony from both the executive branch of the government and some of the persons who claim to have seen UFO's.

"I enclose material which I think will be helpful to you in assessing the advisability of an investigation of UFO's.

"May I first call to your attention a column by Roscoe Drummond, published last Sunday in which Mr. Drummond says, 'Maybe all of these reported sightings are whimsical, imaginary or unreal; but we need a more credible and detached appraisal of the evidence than we are getting.'

"Mr. Drummond goes on to state, 'We need to get all the data drawn together to one place and examined far more objectively than anyone has done so far. A stable public opinion will come from a trustworthy look at the evidence, not from belittling it.'

"'The time has come for the President or Congress to name an objective and respected panel to investigate, appraise, and report on all present and future evidence about what is going on.'

"I agree fully with Mr. Drummond's statements. I also suggest you scan the enclosed series of six articles by Bulkley Griffin of the Griffin-Larrabee News Bureau here. In the last of his articles, published last January, Mr. Griffin says, 'A main conclusion can be briefly stated. It is that the Air Force is misleading the public by its continuing campaign to produce and maintain belief that all sightings can be explained away as misidentification of familiar objects, such as balloons, stars, and aircraft.'

"I have just today received a number of telegrams urging a congressional investigation of UFO's. One is from retired Air Force Col. Harold R.

Brown, Ardmore, Tennessee, who says, 'I have seen UFO. Will be available to testify.'

"Another, from Mrs. Ethyle M. Davis, Eugene, Oregon, reads, 'Nine out of ten people want truth of UFO's Press your investigation to the fullest.'

"Ronald Colier of Los Angeles who identifies himself as 'a scientist from M.I.T.,' urges that you 'do everything in your power to make Air Force Project Blue Book (the AF name for its study and verdicts on UFO reports) known to the people.'

"Are we to assume that everyone who says he has seen UFO's is an unreliable witness?

"A UPI story out of Ann Arbor, Michigan, dated March 21, 1966, states that 'at least 40 persons, including 12 policemen, said today that they saw a strange flying object guarded by four sister ships land in a swamp near here Sunday night.'

"Matt Surrell of Station WJR, Detroit, cites an eye witness account of a recent UFO sighting by Emile Grenier of Ann Arbor, an aeronautical engineer employed by Ford Motor Company. He points out that an aeronautical engineer can hardly be considered an untrustworthy witness.

"In the firm belief that the American public deserves a better explanation than that thus far given by the Air Force, I strongly recommend that there be a committee of the UFO phenomena.

"I think we owe it to the people to establish credibility regarding UFO's and to produce the greatest possible enlightenment on this subject.

Signed

Gerald R. Ford, M.C."[24]

The House did take up Gerald Ford's request and it was the first time, but not the last, that the House ever heard testimony about UFOs, and all of this was right before the eyes of President Lyndon B. Johnson, who was already struggling with the prospects of a war in Vietnam, had based his demands for a War Powers Act on a phony story of an attack by the Vietnamese on the USS *Maddox* and *Turner Joy* in the Gulf of Tonkin, was facing race riots at home in American cities like New York and Los Angeles, and was still anguished with guilt over his complicity in the assassination of President John F. Kennedy. But just about one year after the Hillsdale flap and almost twenty years after the Roswell crash, the Johnson administration would be troubled by another major UFO incident, this one directly threatening national security, when a UFO hovered over Malmstrom Air Force Base in Great Falls, Cascade County, Montana, in March 1967, right before the eyes of the missile launch command of Echo Flight.

UFOs Over Malmstrom Air Force Base

Our Minuteman ICBMs, one of the legs of the US nuclear defense triad, are housed in blast-hardened silos underground, monitored and controlled by launch officers who, at least fifty years ago, communicated with their weapons via a hard-wired circuitry. For anything or anyone to interfere with the launch control mechanisms the hard wiring itself would have to be compromised, and, presumably, that would have to take place by direct physical intervention. But an event at a missile launch facility in rural Montana would challenge that assumption and demonstrate to the Department of Defense that there is a power with the ability to control our most secure weapons, rendering them inoperable. This is that story, which most surely reached the desk of the commander in chief at the time, President Lyndon Johnson.

As documented by former USAF Captain and Minuteman missile launch officer Robert Salas,[25] who's appeared on many talk shows to describe the incident, early in the morning of March 16, 1967, there were reports from security officers patrolling above ground that they had seen unidentified objects hovering above the facility. At least one of the unidentifieds was hovering directly over the Echo Flight silo. Robert Salas said that one member of the security detail who saw the light hovering over the silo was so unnerved by what he saw that "he never again returned to missile security duty."[26]

Later that morning, there was an alert that one of the missiles had gone off line, meaning that it was inoperable. As an officer in charge began calling around the missile site command facilities to determine if there was an existing maintenance problem, other silos began reporting that their missiles were inoperable. As Salas recalled the incident, "Within seconds, the entire flight of ten ICBMs was down! All of their missiles reported a 'No-Go' condition. One by one across the board, each missile had became inoperable. When the checklist procedure had been completed for each missile site, it was discovered that each of the missiles had gone off alert status due to a Guidance and Control (G&C) System fault. Power had not been lost to the sites; the missiles simply were not operational because, for some unexplainable reason, each of their guidance and control systems had malfunctioned."[27]

Captain Salas himself was the deputy missile combat crew commander working below ground at Oscar Flight that morning and a firsthand witness to these events. In his own words he described the sighting of the UFOs over his facility when the reports came in that UFOs were hovering just over the silos. "It was one of those airmen who first saw what at first appeared to be a star begin to zigzag across the sky. Then he saw another light do the same thing, and this time it was larger and closer. He asked his Flight Security Controller (FSC, the Non-Commissioned Officer (NCO) in charge of Launch Control Center site security) to come and take a look. They both stood there watching the lights streak directly above them, stop, change directions at high speed and return overhead. The NCO ran into the building and phoned me at my station in the underground capsule. He reported to me that they had been seeing lights making strange maneuvers over the facility, and that they weren't aircraft. I replied: 'Great. You just keep watching them and let me know if they get any closer.'"[28]

The lights did come closer, triggering the following conversation after a "clearly frightened" noncommissioned officer called in his report:

"Sir, there's one hovering outside the front gate!"

"One what?"

"A UFO! It's just sitting there. We're all just looking at it. What do you want us to do?"

"What? What does it look like?"

"I can't really describe it. It's glowing red. What are we supposed to do?"

"Make sure the site is secure and I'll phone the Command Post."

"Sir, I have to go now, one of the guys just got injured."

When Captain Salas went topside to interview his security noncom, the airman described the UFO his team saw as having a "red glow and appeared to be saucer shaped and he repeated that had been immediately outside the front gate."[29]

Of course, there was a complete investigation of all possible maintenance and operational issues, but Captain Salas remembers that even though maintenance consultants at Boeing were called in to review what had happened and run full checks on all the systems, they found no direct cause for the problem. "No positive cause for the shutdowns was ever found, despite extensive and concentrated effort. One conclusion was that the only way a pulse or noise could be sent in from outside the shielded system was through an electromagnetic pulse (EMP) from an unknown source. The technology of the day made generating an EMP of sufficient magnitude to enter the shielded system a very difficult proposition, requiring large, heavy, bulky equipment. The source of the actual pulse that caused the missile shutdowns remains a mystery to this day."[30]

But there was no heavy equipment present, no specific triggering of a conventional electromagnetic pulse, and nothing that would otherwise impede the current flow because a hard-wired circuit is not easy to interrupt. The inspectors from Boeing also could find no external cause for the event, with Bob Salas writing, "William Dutton, another Boeing Company engineer, checked commercial power interruptions and transients, and stated: 'No anomalies were found in this area.'[31] With no explanation except the presence of multiple UFOs hovering over the site, the government decided to dismiss the event. But, in reality, what was the real message sent to the Pentagon's missile command and ultimately to President Johnson, who was committing US troops to a ground war in southeast Asia, pushing against the paper-thin envelope that could trigger a war with China and possibly the Soviet Union? If that message came from the UFOs over Malmstrom, it was, "be very careful and very afraid. We control your missiles."

Thus, we may believe, it was almost with a sense of relief that Johnson told the nation in 1968 that he would not stand for reelection to the presidency, nor would he accept the nomination of the Democratic Party.

Cue former vice president Richard M. Nixon.

Chapter 12

NIXON, "THE GREAT ONE," AND THE ET IN FLORIDA

President Eisenhower's vice president and, after his loss to Kennedy in the 1960 election, an unsuccessful candidate for governor of California, losing to Pat Brown, father of California's current governor Jerry Brown, Richard Nixon finally won the presidency in 1968, defeating Vice President Hubert Humphrey on an "I have a plan to end the Vietnam War" promise.

He didn't.

But Nixon's presidency, although remembered today as the presidency of Watergate and Nixon's resignation, was also the presidency of our first flights to the moon, the moon landing, the end of Project Blue Book after the Condon Report, the murder of Dr. James McDonald during Nixon's famed "COINTELPRO" operation, and the Nixon/Jackie Gleason ET visit in Florida. We also have reports of the ghost of Richard Nixon himself hovering over his own gravesite in California and a statement by his daughter, Julie Nixon Eisenhower, on the old "Colbert Report" that there exist ghosts in the White House.

But, to be fair, it was the also the presidency that opened the door to our relations with mainland China, amended Truman's 1948 Clean Water Act in 1972, was responsible for the 1970 Clean Air Act, and heeded the advice of Nixon's advisory board to push through the enactment of the Environmental Protection Agency. Despite his manifest personal failings and phobias, attorney Richard Nixon was an institutionalist at heart. And even though he struggled mightily against the pressure of our institutions at the end— and lost—he still believed throughout his career, and came to realize at the end, that the United States was a government of institutions—framed that way and built that way—and not a government of capricious human beings.

Nixon and UFOs

Under Nixon's presidency we reached, orbited, and walked upon the surface the moon while almost at the same time the air force was exploiting a government asset, Dean Edward Condon, to give the air force scientific cover for its decision to terminate Project Blue Book. Although the Condon Committee, as it was called, at the University of Colorado, had been contracted by the air force during the Johnson administration to write a report based on its "independent" study of reported UFO phenomena, a report that would dismiss the question of UFOs as unworthy of study in response to Gerald Ford's letter to Mendel Rivers asking Congress to investigate UFOs, the actual termination of Blue Book took place on December 17, 1969, just about six months after Apollo 11 landed on the moon. Strange juxtaposition of time, especially when we add the third element, the testimony of Professor James McDonald before the Condon Committee, his rejection of their conclusions, and his suicide/murder during Nixon's counterintelligence program to subvert what Nixon believed were his left-wing opponents. You cannot make this up.

Dean Edward Condon, the Air Force, and the Condon Report

1969 saw the end of Project Blue Book with the air force claiming that mostly all of the citizen UFO sightings could be explained away conventionally—a plane, a bird, "swamp gas"—and the military or official sightings as either faulty radar, a temperature inversion, or the moon's being mistaken for a flying craft. There were more than a few "unknowns," however, incidents where the air force or its investigators couldn't come up with a ready explanation, as tortured or implausible as it might have been.

But the notion that an independent scientific investigation of some of our nation's highest profile cases could render an honestly objective opinion was simply false. The first and foremost issue was Dean Edward Condon himself. At a time when President Nixon and his federal agencies believed that the left-wing protesters opposing him were as vicious and as wily as the Communists and posed an existential threat to American democracy, and while at the same time, human beings were reaching the moon, Nixon, who believed in the existence of UFOs and extraterrestrials, agreed to squelch the UFO issue with the release of the Condon Report. Back in the bad old days of the Black List, Nixon's mentor, Senator Joe McCarthy's Cassius, Donald Trump's and Roger Stone's 先生,[1] and soon-to-be mob defense attorney the late Roy Cohn, as was his wont, took a harsh and ugly look at Edward Condon, who, as a physicist, was accused of being a Communist revolutionary sympathizer because he was an adherent of a "revolutionary" physical theory called quantum mechanics. Yes, it was a Luddite argument to equate quantum theory with revolutionary Communism, but it was convenient enough for Cohn's purposes, and it cast a dark stain upon Condon's career even though he was an important physicist and academician. Thus, to purge himself of a Communist taint, Condon was afforded the opportunity of allowing the air force to get out from under the public UFO controversy by putting the kibosh on the reality of UFOs and their importance to the world of science.

The air force made itself clear to Condon: allow us to get out of the UFO business. And there were members of the Condon committee who agreed to go along with what the air force wanted, releasing at least one memo indicating that the committee would not find UFOs cases credible at all and that the air force should be assured that the committee would go along with the air force wished. The committee itself followed along, Edward Condon writing a summary in which he said that the UFO question could effectively be put to rest because there was no science there nor was there a threat to national security.[2]

The Condon Report and the controversy it stirred up was certainly noteworthy as Nixon asked for "four more years" as turmoil in the United States grew in the run-up to the 1972 campaigns. Nixon had spurred our getting to the moon, to fulfilling the desires of a martyred president, and to fighting those "bums" out in the streets and, in the estimation of some, had gotten himself and his military out from under UFOs. Now it was time to enable J. Edgar Hoover to strike back at the one person whose pursuits of UFOs, and the Soviets, could threaten all of that, Professor James McDonald.

Professor James McDonald and COINTELPRO

In his private life, James McDonald had a deep secret he had committed to his journal. He was suicidal. He had testified before the Condon

Commission, he had publicly submitted testimony to Congress in 1968, and to make matters even worse, he had contacted the Soviet consulate in New York to suggest working through their files, were they to have had any, on UFOs to corroborate cases he had investigated. But the FBI believed that the McDonalds had "new left tendencies" and Hoover's FBI took further interest in otherwise "innocuous" UFO queries. The FBI wrote:

"It appears that Professor MC DONALD's letter to [redacted1: probably the Soviet representative at the UN] (contents of which are not known) might have been in itself an innocuous, sincere contact on the part of Professor MC DONALD; however, in view of his background and [redacted2] long time leadership in 'New Left' activities in the Tucson area, it is quite probable that Professor MC DONALD would be highly susceptible to an approach made by a Soviet intelligence, particularly concerning a research done in his field, of atmospheric physics."[3]

As worrisome as all of this might have been for a Cold Warrior like Hoover, when McDonald's wife, Betsy McDonald, housed members of the Black Panther Party at the McDonald home,[4] in the age of COINTELPRO this was the final straw, and Hoover's forces went on alert as can be seen from the FBI file excerpts.[5]

"In October 1968, [redacted: obviously Betsy McDonald] reportedly attended the SDS National Convention in Boulder, Colorado, and was overheard telling a group in Tucson upon her return that she favored violence, if necessary, to achieve the goals of the SDS."

How to eliminate the threat? Inasmuch as Betsy McDonald had shown extreme interest in one of the participants in a local Socialist Party group whose meetings she had been attending, plotters against James McDonald believed that his wife would eventually fall in love with her socialist friend and move to divorce her husband. And that's exactly what happened, resulting in what the FBI knew would take place because they had already broken into McDonald's home and read his journal. He was suicidal, and would in all likelihood attempt suicide. Which he did.

McDonald might have been a brilliant physicist and tireless researcher, but he was a lousy shot. He put his pistol to his head, fired a round, and severed his own optic nerves, thus leaving him still alive, but irreversibly blind. Betsy, his wife, agreed to give up her lover and stay loyal to her disabled husband. But McDonald himself wasn't finished and on the next suicide attempt, he was successful. Thus, the McDonald threat, probably the most important scientist ever to raise questions about the nature of UFOs and how the government dealt with them, ceased to be a threat on Richard Nixon's watch. But that didn't end Nixon's involvement in UFOs.

The Apollo Program

One of the great hallmarks of Richard Nixon's presidency was NASA's Apollo program and the first walk upon the lunar surface. While conspiracy theories are rampant to this day about whether we actually reached the moon or not or whether the moon landing was actually a secret program with a cover video made by producer Stanley Kubrick, the deeper mysteries lie in the strange NASA transmissions from the Apollo astronauts themselves about anomalies they witnessed on the moon. For example, during Apollo 11, when astronaut Neil Armstrong, the first human being to walk upon the lunar surface, reported that he believed that extraterrestrials had been to the moon before humans, had established a base on the moon, and that those ETs were warning us not to build the moon base that had been proposed as early as 1956, during the Eisenhower administration.[6] Both Armstrong and Buzz Aldrin were said, in unconfirmed reports, to have seen UFOs on the moon shortly after their landing in July 1969. Moreover, in one of the early uncensored NASA transmissions, viewers following the Apollo astronauts' adventures heard them talking about a "light" in a crater that seemed to be of unnatural or artificial origin. Although the rest of the transmission was never broadcast by NASA, ham radio operators were said to have heard the following:

NASA: What's there? Mission Control calling Apollo.

Apollo 11: These babies are huge, sir. Enormous! Oh my God. You wouldn't believe it. I'm telling you there are other spacecraft out there, lined up on the far side of the crater edge. They're on the moon, watching us.

In a subsequent interview, Armstrong said, "It was incredible. Of course, we had always known there was a possibility. The fact is, we were warned off. There was never any question then of a space station or a moon city."

The interviewer asked Armstrong how he knew that humans were warned off the moon, something the Chinese and the Indians might soon find out. Armstrong answered, "I can't go into details except to say that their ships were far superior to ours, both in size and technology. Boy were they big, and menacing. There was no question of a space station." He was likely referring to a possible moon base even though there are numerous reports of UFOs buzzing, and caught on camera, at the International Space Station.

Although the story never made it into the American press, Soviet scientists were heavily invested in covering it, with one scientist revealing that Buzz Aldrin took color film of the UFOs from inside the lunar module and continued filming them after he and Armstrong went outside. Years later, Armstrong confirmed that this story was true, but that he couldn't talk about it because the CIA wanted the story covered up.

In 1971, still in Richard Nixon's first term, astronauts on *Apollo 15* had a stunning conversation with NASA mission control about artifacts they discovered on the lunar surface, particularly tracks that they said could only be described as artificial.

> **Astronaut Irwin:** Tracks here as we go down the slope.
> **Cap com:** Just follow the tracks, huh?
> **Irwin:** We know that's a fairly good run. We're bearing 320, hitting 350 on rage for 413 . . . I can't get over those lineations, that layering on Mount Hadley.
> **Scott:** I can't either. That's really spectacular.
> **Irwin:** They sure look beautiful.
> **Scott:** Talk about organization.
> **Irwin:** That's the most organized structure I've ever seen.
> **Scott:** It's so uniform in width.
> **Irwin:** Nothing we've seen before this has shown such uniform thickness from the top of the tracks to the bottom.

These astronauts are talking about artificial tracks on the lunar surface. Who made them? Did NASA report these findings up the bureaucratic chain of command to the White House? Was this one of the startling revelations that prompted President Nixon to reveal the truth about UFOs as far as he could by showing Jackie Gleason alien artifacts and the alien body?[7]

The range of undisclosed NASA transmissions, from the earliest Mercury and Gemini flights to goings on at the ISS, are too long for this book, but even the most cursory studies of the history of NASA's undisclosed files reveal a startling history that has not been revealed to the American public, a history that began during the Eisenhower administration and continues to this very day.

President Nixon Shows Jackie Gleason Extraterrestrials

In a story oft repeated throughout Hollywood and studio friends of film and television star Jackie Gleason, "The Great One" was also friends with Richard Nixon, his golfing buddy, and other presidential friends including Bebe Rebozo. But Gleason, despite his success on television and in more than one poignant motion picture performance such as *The Hustler*, opposite Paul Newman, *Soldier in the Rain*, opposite Steve McQueen, and *Nothing in Common*, opposite Tom Hanks, was yearning for one thing, the truth about UFOS and the paranormal. Gleason had read and collected over fifteen hundred books on UFOs and the paranormal.[8] According to his biographer, William A. Henry, "Jackie Gleason had a lifelong fascination with the supernatural. Everything that Shirley MacLaine was to explore in her exotic life and best-selling book had already been explored by Gleason . . . He would spend small fortunes on everything from financing psychic research to buying a sealed box said to contain actual ectoplasm, the spirit of life itself. He would contact everyone from back-alley charlatans to serious researchers like J. B. Rhine of Duke University and, disdaining the elitism of the scholarly apparatus, would treat them all much the same way."[9] And who better for Gleason to ask about the subject of UFOs than his good friend and Florida neighbor the president of the United States, Richard Nixon. But Nixon, who obviously, despite his repeated denials, knew the truth about UFOs, said he neither could nor would tell him anything substantive about UFOs until, finally, one night an unescorted President Nixon showed up at Jackie Gleason's front door and invited him to take a ride with him. Gleason did not see any Secret Service agents in sight.

Gleason said, as reported by author Larry Warren, who, with Peter Robbins, wrote *Left at East Gate*, and on the Presidential UFO site also said, that the two of them drove to Homestead Air Force Base, where a shocked gate sentry let them in. They drove to a far end of the base to a section that was separated from the rest of the facility—"segregated base" was the term that Warren said Gleason used—where they saw debris from flying saucers in glass cases and then, in Warren's words quoting Jackie Gleason, "Next, we went into an inner chamber and there were six or eight of what looked like glass-topped Coke freezers. Inside them were the mangled remains of what I took to be children. Then—upon closer examination—I saw that some of the other figures looked quite old. Most of them were terribly mangled as if they had been in an accident."[10]

If you've ever wondered whether Nixon knew, about UFOs, that is, the answer is not only a full-throated yes, but he knew where the bodies of ETs were stored. Moreover, lest one think that the Larry Warren story is suspicious, others have told the same story, including Gleason's friends at Sony Pictures when he was making his final film, *Nothing in Common*. Execs at the studio remembered that Gleason would not stop talking about the UFO debris he saw and the alien bodies and how that affected his entire life from that point on. As quoted by his then spouse, Beverly Gleason, to historian Grant Cameron about a book she had planned to write about Jackie, "Jackie had been out very late one night I did not know who he was with. He told me where he was that same evening, he said he had been in South Florida with President Nixon to see some dead aliens there and I believed him, he was very convincing."[11]

Despite denials from members of the Nixon administration, especially his vice president Spiro Agnew, Nixon was, indeed, an avid student of UFOs, knew where they were stored, but nevertheless steadfastly kept one of the military's and the intelligence services' most closely guarded secrets: UFOs were real and we knew all about them.

McMinnville, Oregon, May, 1950, photo taken by
Paul Trent of a UFO over his farm. This photo has
been analyzed, debunked, and undebunked over
and over again. Was it a real craft or a hoax?

Battle of Los Angeles, February, 1942. Searchlights pinpoint targets
in the sky over the California coast.

Army Air Force major Jesse Marcel, intelligence officer at the 509th in Roswell, New Mexico, who led a retrieval unit to collect the debris of a crashed UFO outside of Roswell. In 1978, he revealed the full truth of what happened at Roswell to UFO researcher Stanton Friedman.

Army Lieutenant Colonel Philip J. Corso in a 1997 photo. Col. Corso wrote that he first saw the Roswell debris at Fort Riley in 1947, then again at the Pentagon in 1961. Photo courtesy of Nancy Birnes.

Photo by Rex Heflin of what he claimed to be a UFO
over a road in Orange County, California. Photo was taken
through his front windshield.

Heflin's second photo as the object passed across the
hood of his pickup truck.

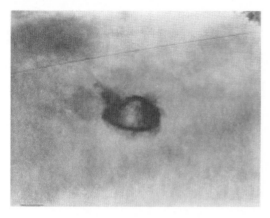

Smoke ring left by the UFO taken from outside the cab of Heflin's his truck. Notice the difference in the sky.

Rex Heflin describes how the UFO looked to him. See his interview on Youtube

The Hills, Betty and Barney, whose reports of an 1961 UFO abduction on a lonely road in New Hampshire became a cover story in Look Magazine and caused a nation-wide sensation about ETs and their interactions with humans.

UFOs over the nation's capitol in July, 1952 during the "summer of the saucers."

Marion "Black Mac" Magruder, who saw an alien at Wright Field in 1948 as part of the National Air War College class of 1948. Mac Magruder was the first Marine Air Squadron leader to use radar-vectored night fighting tactics at the Battle of Okinawa in World War II. Photo Courtesy of *UFO Magazine.*

Ed and Frances Walters of Gulf Breeze, Florida. Ed Walters's photos and stories of his encounters with UFOs over his home and his neighborhood stunned the media in the late 1980s.

Close up of an Ed Walters photo of a UFO.

Enceladus, the largest moon of Saturn, which has shown evidence of subsurface oceans, surface plumes of water, and the possibility of organic materials in its atmosphere. It's one of the candidates for extraterrestrial life in our solar system. Photo courtesy of NASA.

Artist conception of the Phoenix Lights from March,
1997. Courtesy of Nancy Birnes.

Svalbard seed vault in Spitsbergen Norway, where the world's repository of seeds sits in a deep freeze, but which has already been compromised due to the melting of the ice caps and climate change.

Chapter 13

PRESIDENT GERALD FORD'S QUEST FOR UFOS

Gerald Ford's UFOs

Even before he assumed the presidency, as a member of Congress Gerald Ford had already run afoul of the head of Project Blue Book, Air Force Lieutenant Colonel Hector Quintanilla, who blasted Ford for asking Congress to engage in a "circus" because of his request to have Congress take up the question of UFOs. Even President Johnson's Secretary of Defense Robert McNamara was of two minds about Ford's request to Congress after the 1966 Hillsdale, Michigan, sightings, saying that he doubted there was any real substance to UFO reports but conceding that there might be something there, a nuanced approach which did not get him in trouble with anybody except for the head of Project Blue Book, not an enemy you want to make. Lieutenant Colonel Quintanilla wrote that Ford's request for Congressional hearings on UFOs was only a "political circus" and that he "didn't want any part in it."[1]

But, though admonished in public and then warned, Ford persisted. He told Major George Filer in response to a letter Filer had written him that he did not pursue the UFO question and that those people in government he wrote to about it refused to give him any information about it.[2] Nevertheless, Ford had indicated on more than one occasion that he was open to finding out what the military knew about the phenomena. With regard to the government's keeping UFOs a secret subject, Ford stated that if there were presidents who had been given those secret files, he was not one of them and could not shed any light on the subject.

Back in the middle seventies, however, when Ford was serving out the remainder of Nixon's term, he did talk about UFOs, indicating that he was interested in finding out the truth regarding their existence in our skies and the interactions witnesses had with them. He repeated his intention to ferret out whatever files existed. Ford's interest in revealing the truth about UFOs made some minor news at the time in early 1975, after he had taken office upon the resignation of Richard Nixon. Although Ford's statement of his intention did not make national headlines, a few radio shows in affiliate local markets picked up the story. In New York, the story received some traction with listeners phoning in to station managers that they wanted to hear more about President Ford and UFOs. Then, upon the heels of the stories about Ford and UFOs, there were two separate assassination attempts upon Ford, both of which were in California and both of which were perpetrated by members of the Charles Manson gang within weeks of each other. These were the only assassination attempts in American history to have been perpetrated by women, one of whom, Sara Jane Moore, was an FBI informant at the time.

The first attempt was made by Lynette Fromme, also known as "Squeaky," on the California State House grounds in Sacramento on September 5, 1975. Ostensibly, the reason for Fromme's attack was to protest against Ford's stance against continuing the environmental protections, specifically against smog in California. Ford had agreed to replace them, and met with then current California governor Jerry Brown at a conference of business leaders. Fromme, one of the original members of the Manson gang, said the attempt was means of protesting Ford's ties to businesses that were polluting California's air. The assassination attempt failed even though Fromme was only a few feet away from the president. It was later explained that as Fromme drew her handgun and aimed it point blank at the president, a Secret Service agent managed to slip his hand between the gun's hammer and the barrel, thereby preventing the hammer from hitting the cartridge and the gun did not fire. Ford's life was spared.

Weeks later on September 17, 1975, this time in San Francisco, Sara Jane Moore, a dissident who had been arrested by the police a day earlier on a gun charge, attempted to assassinate President Ford from over forty feet away. The gun she had initially brought to the crime had been confiscated by law enforcement the previous day, but she purchased a different weapon on the morning of the attempt, a weapon she had not used before. Her first shot missed and when she aimed for her second shot, a member of the crowd nearby knocked her arm and she missed again. Both Fromme and Moore served out their respective prison sentences.

It might seem coincidental or even strangely conspiratorial that two members of the same criminal gang would attempt back-to-back assassinations of the president. But after delving further into their crimes and interviewing Charles Manson himself, as he told psychologist and author the late Joel Norris years later, a new twist to the story comes to light.[3] One of the claims Manson made to Norris during an interview for his story in *Serial Killers* was his revelation that he and his gang worked black ops for the government. In other words, he claimed, he was actually leading a "men in black" enterprise whenever ordered to do so by clandestine forces within the government. Pressed further on why the Manson gang was ordered to strike at Ford, not once but twice, was, Manson only told Norris, "UFOs."[4]

The Parvis Jafari Iranian Air Force Incident Reaches the White House

Toward the last months of the Ford administration and during the final years of the Shah's rule in Iran, an Iranian Air Force squadron leader, Major Parvis Jafari, had a harrowing dogfight with a UFO over the skies of Tehran, an incident that was monitored via radio transmissions, on radar, and by United States Air Force intelligence, which sent a full report to the desk of President Ford and thence to his successor, President Jimmy Carter.

The entire Jafari incident, which he has described on television and to the team at *UFO Hunters*, began when a Soviet MiG was sighted as it crossed into Iranian airspace. An F-4 from Jafari's squadron was vectored by his flight controllers to keep the MiG in view, but not engage it or follow it into Soviet airspace. At the same time, according to a US intelligence report about the incident, a report that was circulated throughout the national security community and wound up on Ford's desk, residents began calling to the Imperial Air Force that they had seen a "strange" object in the sky overhead.[5]

According to the official report, the area commander of the Imperial Air Force, though he believed that the residents were probably looking at a celestial object, ultimately decided to dispatch an F-4 Phantom jet for

a look-see. As the pilot approached the object, he made visual confirmation and said that the object was so brilliant, he could see it from seventy
miles away. As he approached the object to about twenty-five miles the
pilot reported that he lost all instrumentation as well as communications
on radio and internally. The pilot broke off contact with the object, turned
away, and headed back to base, whereupon his instrumentation and communications returned, perhaps, the report speculates, because the Iranian
pilot was no longer a threat to the object.

A second F-4 was vectored toward the object and when the second
pilot's back-seater acquired the object on his radar and the plane closed to
twenty-five miles, the object accelerated away and then maintained a con-
stant distance from the F-4 of twenty-five miles. The size of the radar image
the object made was approximately the size of large fuel tanker even though
the visual size of the object was "difficult to discern" because of the object's
"intense brilliance," which was like "flaming strobe lights arranged in a rectangular pattern and alternating blue, green, red, and orange in color. The
sequence of the lights was so fast that all the colors could be seen at once."[6]

The two craft, the F-4 and the object, continued on course south of
Tehran when a second brilliant object appeared, coming out of the first
object, that looked to be "twice the size of the moon." The second object
headed straight for the F-4 at a "very fast rate of speed," upon which movement the pilot of the F-4 attempted to fire an AIM air-to-air missile. That
pilot was the squadron leader Parvis Jafari. However, at the moment Jafari
attempted to fire the missile, his weapons control panel went off line and
became inoperable. Jafari also lost all communications including his internal phone. Jafari broke off the engagement, diving away from the object,
which followed him, and the second object returned to the first object. Suddenly another object emerged from the "primary object" and dived down
on Jafari at a "great rate of speed," although by this time Jafari had regained
his communications. Jafari and his back-seater thought at first that the diving object would crash into the desert, but it slowed and gradually landed
"gently" on the desert floor where it "cast a very bright light."

Jafari noted where the object had come to rest and returned to base,
whereupon he saw a third flashing object in the distance, this time in the
shape of a cylinder, but his full control panel and communications had
come back on line. After landing back at his base in Mehrabad, Jafari

reported that another crew in a helicopter was sent back out to look for the object that had landed in the desert. It was now full daylight, and the helicopter crew found a small house in the vicinity where the object had landed. They landed and asked the people in the house if they had experienced anything strange earlier that morning in the darkness, and were told that there was a bright flash just like lightning where the object was believed to have landed. The area was analyzed for any traces of radiation, but the report does not cite the results of that analysis. The circulation list for this Department of Defense report specifically lists "the White House, Washington, DC" as a destination, which means that as of September 1976, during Ford's last months in office while he was running for reelection, he would have been presented with this stunning report of a UFO dogfight over the Iranian desert.

Parvis Jafari told his version of the encounter, which can be seen at the Black Vault Files, a story he has told repeatedly to reporters and in documentaries about UFOs.[7] In Jafari's version, he was dispatched to lead one of his fighters back to base after the F-4 had spotted a Soviet MiG entering Iranian airspace. After Jafari ordered the pilot not to engage and return to base, that's when he spotted the brilliant object. When he closed on it, the object moved away. He followed it, whereupon it turned towards him in a manner that seemed threatening. Jafari attempted to get a radar lock on the object to acquire his target when his missile control panel went dead and his radio went out. At that point he noticed that his flight controls were becoming unresponsive. He broke off, his controls returning to full operability, but the object not only followed him, it seemed to bounce from one spot on the right side of his F-4 to the other side seemingly in an instant, a maneuver no conventional airplane could make and reminiscent of the RB47 case two decades earlier. Jafari said that while this was going on, he experienced a brief period of "missing time," but was able to return to base safely.

Jafari, who ultimately became a member of the chiefs of military staff for the Ayatollah's military during the country's eight-year war with Iraq, was a credible witness who had made numerous trips to the United States where he appeared on different talk shows, including *UFO Hunters*. He said that upon his return to base after his UFO encounter, he was interviewed by uniformed officers in the United States Air Force as well as by

English-speaking agents in civilian clothing. The report that was sent to the White House in September 1976 would have been part of a briefing for the president, Gerald Ford, and more than likely would have also been part of a briefing for the incoming president, Jimmy Carter, after the election and, more than likely, after he assumed office the following January.

Thus, President Ford, who had asked for Congressional hearings after the 1966 Hillsdale, Michigan, incident, who had called for the public release of UFO files, and had been targeted for assassination twice by members of a gang whose incarcerated leader said he was working for the secret government, finally came face to face with the report of a military dogfight with a UFO.

Chapter 14

THE TURBULENT PRESIDENCY OF JIMMY CARTER

The Carter UFO Sighting

Jimmy Carter was no stranger to UFOs and stories about them. In 1969, Carter, along with friends leaving a Lions' Club meeting in his native Georgia while he was the sitting governor of Georgia, saw, and reported, a UFO that flew by. In an interview with Jim Clash in *Forbes* magazine, the Director of the Mutual UFO Network, Jan Harzan, said, "President Carter had his sighting while governor of Georgia in 1969 at a talk he was giving to the Lion's Club in Leary, Georgia. He and a dozen other people witnessed a strong blue light in the sky which he could not explain. The Governor was interviewed by a UFO investigator and his report was filed with his signature on it. One may see this report on the Internet by googling, 'Jimmy Carter UFO report.'¹ Based on personal experience with this UFO, Carter vowed in 1975 that if elected president he would make every file the U.S. Government had on UFOs available to the public. Of course, we know that never happened. Who stopped him? Does Mika know? Isn't the president the most powerful man in the world?"² The report then governor Carter filed said:

1. Name *Jimmy Carter* Place of Employment
    Address *State Capitol Atlanta* Occupation *Governor*
    Date of Birth
    Education *Graduate*
    Special Training *Nuclear Physics*
    Telephone (404) 656-1776 Military Service *U.S. Navy*

2. Date of Observation *October 1969* Time AM PM Time Zone
   7:15 *EST*

3. Locality of Observation *Leary, Georgia*

4. How long did you see the object?
   [BLANK] Hours *10-12* Minutes [BLANK] Seconds

5. Please describe weather conditions and the type of sky; i.e. bright,
   daylight, nighttime, dusk, etc. *Shortly after dark.*

6. Position of the Sun or Moon in relation to the object and to you.
   Not in sight.

7. If seen at night, twilight, or dawn, were the stars or moon visible?
   Stars.

8. Were there more than one object? *No.* If so, please tell how many,
   and draw a sketch of what you saw, indicating direction of movement,
   if any. [BLANK]

9. Please describe the object(s) in detail. For instance, did it (they)
   appear solid, or only as a source of light; was it revolving, etc.? Please
   use additional sheets of paper, if necessary. [BLANK]

10. Was the object(s) brighter than the background of the sky? *Yes.*

11. If so, compare the brightness with the Sun, Moon, headlights, etc.
    At one time, as bright as the moon.

12. Did the object(s)—(Please elaborate, if you can give details.)
    a. Appear to stand still at any time? *yes*
    b. Suddenly speed up and rush away at any time? [BLANK]
    c. Break up into parts or explode? [BLANK]
    d. Give off smoke? [BLANK]
    e. Leave any visible trail? [BLANK]
    f. Drop anything? [BLANK]
    g. Change brightness? *yes*
    h. Change shape? *size*
    i. Change color? *yes*
    *Seemed to move toward us from a distance, stopped—moved
    partially away—returned, then departed. Bluish at first, then reddish,
    luminous, not solid.*

13. Did object(s) at any time pass in front of, or behind of, anything? If
    so, please elaborate giving distance, size, etc., if possible.
    no.

14. Was there any wind? *no.* If so, please give direction and speed. [BLANK]

15. Did you observe the object(s) through an optical instrument or other aid, windshield, windowpane, storm window, screening, etc? What? *no.*

16. Did the object(s) have any sound? *no* What kind? How loud?

17. Please tell if the object(s) was (were)—a. Fuzzy or blurred. [BLANK] b. Like a bright star. [BLANK] c. Sharply outlined. *x*

18. Was the object—a. Self-luminous? *x* b. Dull finish? [BLANK] c. Reflecting? [BLANK] d. Transparent [BLANK]

19. Did the object(s) rise or fall while in motion? *came close, moved away-came close then moved away.*

20. Tell the apparent size of the object(s) when compared with the following held at arm's length:

    a. Pinhead

    b. Pea

    c. Dime

    d. Nickel

    e. Half dollar

    f. Silver dollar

    g. Orange

    h. Grapefruit

    i. Larger

    Or, if easier, give apparent size in inches on a ruler held at arms length. *About the same as moon, maybe a little smaller. Varied from brighter/larger than planet to apparent size of moon.*

21. How did you happen to notice the object(s)? *10-12 men all watched it. Brightness attracted us.*

22. Where were you and what were you doing at the time? *Outdoors waiting for a meeting to begin at 7:30 pm.*

23. How did the object(s) disappear from view? *Moved to distance then disappeared*

24. Compare the speed of the object(s) with a piston or jet aircraft at the same apparent altitude. *Not pertinent*

25. Were there any conventional aircraft in the location at the time or immediately afterwards? If so, please elaborate. *No.*

26. Please estimate the distance of the object(s). *Difficult. Maybe 300-1000 yards.*

27. What was the elevation of the object(s) in the sky? Please mark on this hemisphere sketch. *About 30 [degrees] above horizon.*
   [Mark on arc representing sky elevation looks marked at about 30 degrees.]

28. Names and addresses of other witnesses, if any. *Ten members of Leary Georgia Lions Club.*

29. What do you think you saw?
   a. Extraterrestrial device? [BLANK]
   b. UFO? [BLANK]
   c. Planet or star? [BLANK]
   d. Aircraft? [BLANK]
   e. Satellite? [BLANK]
   f. Hoax? [BLANK]
   g. Other? (Please specify). [BLANK]

30. Please describe your feeling and reactions during the sighting. Were you calm, nervous, frightened, apprehensive, awed, etc.? If you wish your answer to this question to remain confidential, please indicate with a check mark. (User a separate sheet if necessary.) [BLANK]

31. Please draw a map of the locality of the observation showing North; your position; the direction from which the object(s) appeared and disappeared from view; the direction of its course over the area; roads, towns, villages, railroads, and other landmarks within a mile. *Appeared from West—About 30 [degrees] up.*

32. Is there an airport, military, governmental, or research installation in the area? *No*

33. Have you seen other objects of an unidentified nature? If so, please describe these observations, using a separate sheet of paper. *No*

34. Please enclose photographs, motion pictures, news clippings, notes of radio or television programs (include time, station and date, if possible) regarding this or similar observations, or any other background material. We will return the material to you if requested. *None.*

34. Were you interrogated by Air Force investigators? By any other federal, state, county, or local officials? If so, please state the name

and rank or title of the agent, his office, and details as to where and when the questioning took place. [BLANK]

Were you asked or told not to reveal or discuss the incident? If so, were any reasons or official orders mentioned? Please elaborate carefully. *No.*

35   We should like to use your name in connection with this report. This action will encourage other responsible citizens to report similar observations to NICAP. However, if you prefer, we will keep your name confidential. Please note your choice by checking the proper statement below. In any case, please fill in all parts of the form, for our own confidential files. Thank you for your cooperation.

You may use my name (*x*)

Please keep my name confidential ( )

37.   Date of filling out this report Signature: [signed].³

This was a very comprehensive report and the description the governor gave, all the more credible because of his background as a Naval Academy graduate and naval officer, was specific to the point where on the campaign trail he was directly asked whether he would reveal the UFO files the government had. Carter said that he would, and made that a promise.

Briefing by DCI Bush

President-elect Jimmy Carter began the process, he believed, of untangling the UFO story as early as his first briefing with then director of central intelligence George H. W. Bush, where, according to at least one witness named Marcia Smith, Carter asked DCI Bush, "I want to have the information that we have on UFOs and extraterrestrial intelligence. I want to know about this as President."⁴ Reportedly, one of the reasons that the DCI said he could not reveal information to Carter concerned the fact that the "sources and methods" of our intelligence-gathering resources would be revealed, and those were above the president-elect's level of top secret clearance. Thus, he needed to have been read in on a "need to know" basis. But Bush replied that he would not give President-elect Carter the information he requested because he had "no need to know" and the information "that existed" was released on a "need to know basis only." Bush continued, saying, "simple curiosity on the part of the President wasn't adequate."⁵ Nevertheless, DCI

Bush held out a tantalizing clue: some of the information that President-elect Carter wanted revealed was already in the National Archives and readily available for viewing.

This aspect of the story was also revealed to the press by human rights lawyer Daniel Sheehan, who said that back in January 1977, he was led into the basement of the Library of Congress at the behest of the National Research Service, following up on a request from the newly elected President Carter to research UFO files. Carter was following up on Bush's directive to look in the available government files, many of which were still classified, for information about UFOs. In order to facilitate that search request, Sheehan was granted authority to search some of the relevant files from Project Blue Book. Although Blue Book's purpose was the official debunking of UFOs, Sheehan said, "there was in fact a secret portion [of Blue Book material] where they actually sequestered the information that showed genuine UFOs and discussed the problem they were having with these and that's what I got access to. And I saw the unquestioned photographs of UFOs, one of which had crashed into a site where there was snow on the ground and the UFO dug a huge trough through the field and impacted into a dirt embankment, and I was able to see symbols on the base of the dome of the UFO and I actually got to trace those. We have a record of those."[6] Sheehan also said that Bush had asked Carter to keep him on as CIA Director and if Carter agreed to keep him on, "he would make some of this information available to the president." It was that statement, Sheehan said, that motivated President Carter to begin the investigation into UFOs, which he would have revealed after he became president.[7]

On the campaign trail, Jimmy Carter had made a pledge to release the files about UFOs insofar as they did not compromise national security, especially the "sources and methods" that former DCI Bush was referring to in his refusal to release the CIA's information in December 1976, during Carter's pre-inauguration briefings. But once he took office, the story goes, Carter was specifically warned in person that any release of UFO files would certainly compromise national security, and Carter not only backed off his campaign promise but distanced himself from the entire UFO question, recanting his own UFO sighting report and then recanting his previous recantation. It has been a strange turn of events over the years for President Carter with respect to his search for information about UFOs that he could disclose without legal or national security consequences. And who had the authority to walk into the Oval Office to warn President Carter in the bluntest of terms about the relationship between his promised UFO disclosure and the future of his presidency? Was it Dr. Zbigniew Brzezinski, a past president of the Trilateral Commission, who was also part of President Eisenhower's own commission to study and curate the evidence of UFOs? And if it was, what, if anything, did he tell his daughter, Mika, now the MSNBC talk show co-host on *Morning Joe?*

SETI's "WOW" Signal and Carter's "ET Initiative"

"We are extraterrestrials," Search for Extraterrestrial Intelligence (SETI) director and astronomer Dr. Seth Shostak once told the folks on *UFO Hunters.* "Martians, actually, because it is likely that pieces of the planet Mars containing organic chemicals were broken off by meteor impacts early in the solar system's history and landed into Earth's oceans where they seeded the process of life on Earth."[8] True or not, coming from the very person charged with finding extraterrestrial intelligent life in the universe via a scrubbing of distant star systems for coherent intelligence-generated radio signals, this statement was indeed intriguing.

Dr. Shostak has explained that in August 1977, a seemingly unconventional "narrowband" audio signal was picked up by a radio telescope at Ohio State University that was anomalous enough to trigger interest that the telescope might just have picked up an alien broadcast. One of the astronomers working for SETI, Jerry Ehman, was so excited about what the telescope had recorded that when he reviewed the printout, he wrote the word "WOW" on the printout. You can listen to the signal here: https://www.youtube.com/watch?v=iiJLnQtvUOE. The origin of the signal, which lasted for over a minute and seemed to come from the constellation Sagittarius, has never been officially sourced and UFO researchers still hold out hope that it actually was a transmission from an extraterrestrial culture even though other UFO researchers have theorized that the signal could have emanated from a passing comet. However, Shostak explained that in the absence of a repeat of that signal or any positive returns from other radio telescopes scanning that portion of the sky, the WOW signal still remains an anomaly.

Nevertheless, for a president that filed his own UFO sighting report and who was probably very much in the loop concerning the UFO incident in Iran at the end of President Ford's term in office, the possible discovery of a radio signal from an alien culture sparked a renewed interest in a larger question: what would actually happen if human beings did make contact with an alien species? As a result, President Carter initiated an internal project to find out what information the intelligence services had on UFOs, but was rebuffed by the CIA and probably the NSA even though the CIA did admit in a report released almost fifteen years after Carter left office that agency did investigate reports of UFOs starting in 1947.[9]

Other researchers have suggested that Carter even broached a domestic policy agenda regarding contact with extraterrestrial species and even approached the United Nations about developing protocols for ET contact. However, these administration initiatives were also rebuffed not only by the intelligence agencies, but by the Department of Defense, which threatened to pull funds away from any contracted research entity working for them on the grounds that since UFOs did not exist, the DoD would not waste its money on entities engaging in a fruitless search for nonexistent UFOs.[10]

By the end of Carter's term in office, particularly after he had lost the November election to California governor Ronald Reagan, a major UFO story would break at one of the major NATO air bases in the United Kingdom, an air base staffed by the United States Air Force where nuclear warheads were stored in anticipation of a Soviet invasion of Europe and a naval thrust into the North Sea. This was the December 1980 RAF Bentwaters and Rendlesham Forest incident.

The RAF Bentwaters Incident

This is probably one of the best-documented, debunked, then undebunked UFO cases in recent years. And the news of the events at RAF Bentwaters would most certainly have landed right on President Carter's desk because it involved a violation of highly protected British and NATO airspace over a base where nuclear warheads were stored while, at the same time, their presence at the base was officially kept secret from the British people. Also happening at the same time were the Soviet invasion of Afghanistan and the failure of Carter's attempt to free the American hostages held in Tehran.

The story of a UFO encounter over RAF Bentwaters and in Rendlesham Forest in the UK in 1980 has become known as "Britain's Roswell." Like the American Roswell, it was an incident that, because of its high strangeness, became the benchmark of what a UFO encounter is supposed to be. The incident involved:

- An actual UFO landing
- Tripod landing impressions in hard forest soil
- A USAF sergeant, James Penniston, who actually touched the craft and copied in his notebook a set of symbols on the craft's hull
- A wild pursuit of the floating object through the woods led by the RAF Bentwaters deputy base commander, Charles Halt, who recorded the entire chase on his hand-held tape recorder
- A spectacular light show in plowed farm field on the perimeter of the forest that ended when the object, whatever it was, split into five objects and took off
- A mystery witness in an observation tower who had observed the entire event and more

It is arguable that the mystery surrounding the incident might have died away if the United States Air Force had gone to any lengths to explain away what had taken place. But that did not happen. Instead, the entire incident became entangled in a jurisdictional tug of war between the defense and intelligence agencies of two governments. The NATO base at Bentwaters was ostensibly United States territory staffed by American military personnel. However, the event itself took place in adjoining Rendlesham Forest, British territory. Therefore, whose responsibility was it to undertake the investigation, the UK Ministry of Defence or the US Department of Defense? This was one of the initial options dropped in front of President Carter.

Neither agency wanted to handle this hot potato. As a result, the case has never been fully investigated by the authorities even though UFO researchers from both sides of the Atlantic have investigated and extensively written about the case. Accordingly, with all the loose ends dangling in the wind, RAF Bentwaters and the Rendlesham Forest UFO incident has became the subject of many investigations; the case even turned up at the National Press Club in 2007 in Washington, DC.[11]

Of all the speakers at the National Press Club in 2007, the first group comprised the witnesses to the December 1980 incident at RAF Bentwaters and Rendlesham Forest. Two speakers, former USAF Lieutenant Colonel Charles Halt, the deputy commander of RAF Bentwaters, and USAF Sergeant James Penniston, a member of Charles Halt's security detail investigating the strange lights over the base at the end of December 1980, revealed details of the event. In his presentation, Charles Halt described the entire incident and how he became involved when security personnel reported to him that they had witnessed a strange object penetrating restricted air space over the base. James Penniston described how he'd been a part of the security detail accompanying Halt and what he saw in the clearing in the forest, how he touched the strange object that had landed, and the writing or lettering he saw on the side of it.

The incident, as Halt and Penniston described it, defied conventional attempts to explain it. Yet, over the years, skeptics had come out of the woodwork to assert many conventional explanations for what the United States Air Force personnel say they witnessed, explanations which themselves seemed even more implausible than the possibility that the security detail on two successive nights had witnessed an encounter with an unearthly craft.

The Bentwaters incident took place over different nights so that depending upon to whom you're talking, you're getting a different perspective. Ultimately, the main through-line of the story was the experience of deputy base commander Charles Halt because he recorded a running commentary of the entire event on the third night on his voice recorder. Nevertheless, the events began even before Charles Halt became involved when security personnel at the key NATO base on the North Sea responded to reports of strange lights buzzing overhead and a possible landing in Rendlesham Forest, a densely wooded area that separates RAF Bentwaters from RAF Woodbridge.

The first sightings took place at 3 a.m. on December 26, 1980, when US Air Force security spotted lights over the East Gate at RAF Bentwaters. At first, according to Larry Warren, writing with Peter Robbins in *Left At East Gate*, the security team thought it was an aircraft that had gone down, and a team went to investigate. But it was not a downed aircraft at all. Leaving the base and venturing into Rendlesham Forest, which was equivalent to leaving United States territory and entering the United Kingdom, the security team saw a series of lights in the sky and moving through the forest. They came upon a glowing object, roughly the shape of an egg or a cone, hovering just off the ground. According to members of the team, the object then touched down, leaving impressions from its triangular shaped landing gear in the wet sand. Air force sergeant Jim Penniston, part of the security detail, said that he actually touched the object and made sketches of lettering he said he saw along the object's body.

Later that morning, just after daybreak, the servicemen returned to the spot in the forest where the object was reported to have landed and saw three landing impressions in the dense sand. Jim Penniston took plaster casts of the impressions, casts which he displayed to the audience at the National Press Club Disclosure conference in November 2007. Penniston also posted on the Internet his copy of the graphics he said he saw on the object. Later that morning, local British police were summoned back to the site to see the landing impressions. The police had been there before dawn investigating the lights and believed them to be from the Orford Ness lighthouse nearby on the coast of the North Sea. The police were reported to have said that the landing impressions in the sand were the marks of small animals. They were "rabbit scrapings," a local resident had said, which still today makes no sense because rabbits tend to dig into the ground at an angle and these impressions were relatively shallow and were straight down as if they resulted from compression, not digging.

Colonel Halt said that news of the strange events had circulated through the command structure when, at a post-Christmas and pre-New Year's party, he was notified that the strange lights were back.[12] Halt was a skeptic and believed there was nothing to the strange lights story. He assured his boss, wing commander Gordon Williams, that he would investigate this incident himself and promised to "get to the bottom" of whatever it was. Although he felt at first that this was much ado about nothing, he said it was his intention to go out there and "put it all to rest."[13] What he found out, however, would effectively change his life.

Seeking to investigate the reports that the lights were back, Colonel Halt led a security team through the east gate of the base into Rendlesham Forest, where, technically, the United States Air Force had no authority. But any threat to the perimeter of the base needed to be investigated, and that was Colonel Halt's job that night. In addition to his security detail, Halt had ordered up powerful lights running on their own generators to illuminate the area outside the base and, of course, a communications system to keep the different security details in contact with each other.

Charles Halt was, and still is, a very meticulous person. On December 27, 1980, the night he led his team into the forest, Colonel Halt brought along his Lanier voice recorder to preserve a verbal memo of any incidents the security detail might witness so he could include all of them in a report he knew he would write even though there was probably nothing there, which he would deliver to his base commander. But he also kept an open mind and figured that if his trusted security personnel reported something, there might be something to investigate even if turned out not to be a UFO. He never expected to encounter what the team observed as they made their way into the forest.

On Colonel Halt's voice recording, he describes a light playing through the trees as his team entered Rendlesham Forest. Halt later said in his presentation at the November 2007 National Press Club conference that the lights he saw seemed to have affected the radio transmitters as they intermittently cut out, causing some confusion among the members of the security detail. Colonel Halt said that he and his team followed the lights deeper into the forest where the rounded triangular craft that had been observed the night before led his men through the foliage and to the edge of a clearing.

Sergeant James Penniston, who was part of the detail the night before, had said in his presentation about that night that when his team entered the forest to observe the lights, the object descended through the trees and touched down on the forest floor, its three landing pads making a triangular impression in the cold dirt. While the team watched the object in awe, Sergeant James Penniston carefully took note of its size and shape. As he told the gathering at the Washington Press Club, it was a "triangular craft" about nine feet long and six and a half feet high that had blue and yellow lights swirling around its surface.

"We started experiencing radio difficulties," Penniston said. "The air around us was electrically charged and we could feel it on our clothes, our skin, and our hair." Penniston approached it and laid his hands upon its surface. Although it was glowing brightly, the object, Penniston said, was cold to the touch. Penniston also noticed that there were designs, hieroglyphics almost, along the side of the object, the largest of which in the center was a triangle. He sketched these in a book he had brought along to record the events of the evening.[14]

After what seemed to be almost forty-five minutes, the bright object lifted off the forest floor and shot away through the trees in "the blink of an eye," Penniston said. "And over eighty air force personnel witnessed the takeoff."

On the second night, the triangular object that led the pursuing security detail to the edge of the clearing touched down on the ground as if it were waiting. As Halt said on the History Channel,[15] the object that landed in the clearing was so bright that its light reflected off the windows of a farm house at the other edge of the clearing, making the house look as if it were on fire. The air force detail didn't approach the object again because the object was now on private property, a farm, that was off limits to air force personnel absent permission from the landowner.

While Halt and his men watched the object from the edge of the forest, it seemed to grow brighter and suddenly split into five different lights, all of which suddenly took off and could be seen shooting over the tree tops and into the sky. The men were still awestruck by the entire sequence of events. Some members of the detail were disoriented by the incident but others were fully alert and wanted to report what had happened.

Halt kept a copy of his voice recording and wrote out a full report of the incident. He later swore out an affidavit in which he said that he believed what he and his men saw was an actual extraterrestrial object. He wrote:

"I believe the objects that I saw at close quarter were extraterrestrial in origin and that the security services of both the United States and the United Kingdom have attempted—both then and now—to subvert the significance of what occurred at Rendlesham Forest and RAF Bentwaters by the use of well-practiced methods of disinformation."

Halt was referring to the fact that in the aftermath of the incident, both the United States Air Force and the British Ministry of Defence were given copes of Halt's report and also eyewitness accounts of the entire incident. They also had access to James Penniston's sketch of the graphics on the object, but neither defense agency did anything. What made matters worse was that in the aftermath of the incident, investigators took radiation readings from the object's landing site and found that the radiation at the site was many times higher than the background radiation. And Penniston himself took plaster casts of the impressions made by the landing gear of the craft, casts that he showed at the 2007 National Press Club conference. In interviews Penniston has said that the light was no illusion, nor was it the light from a nearby lighthouse. It was, he said, "definitely mechanical in nature."

In summary, the object, in addition to having been witnessed by multiple professional service personnel, also left palpable and measurable trace evidence on the ground. But still neither the British nor the Americans pursued any official inquiry into the incident. And this is what made Colonel Halt frustrated at the official lack of response.

The initial release of information from former USAF personnel prompted a flurry of skeptical responses. Most of the responses centered on the nature of the light, claiming that the air force personnel mistook the beacon from the Orford Ness lighthouse on the coast of the North Sea as the light that led them through the forest. Others interviewed residents who lived by the base who claimed that the Americans mistook the ground impressions for "rabbit scrapings." However, at least one of the residents, who was adamant about what he said was the Americans' lack of understanding about what they actually saw, also admitted that days after the incident he had been visited by plainclothes investigators from an unnamed British agency, who interrogated him about the incident. Who were these men asking about what had happened in Rendlesham Forest even before Colonel Halt had filed his report? How did they know there was an incident in the forest? How did they arrive on the scene so quickly?

One wonders whether the sudden appearance of these men and their immediate response to an incident that hadn't actually been reported yet was more of an intimidation than an information-gathering visit. The witness in question, one Vincent Thurkettle, admitted that he talked to the investigators and told them that he didn't believe the American service personnel saw anything anomalous. He said he believed they saw the light from the Orford Ness lighthouse and mistook holes that rabbits had dug for the impressions of the landing craft. Colonel Halt rejected Thurkettle's characterization of the light in the forest and the skeptics' claims that the Americans mistook what they saw. Halt said:

"While in Rendlesham Forest, our security team observed a light that looked like a large eye, red in color, moving through the trees. After a

few minutes this object began dripping something that looked like mol-
ten metal. A short while later it broke into several smaller, white-colored
objects, which flew away in all directions. Claims by skeptics that this was
merely a sweeping beam from a distant lighthouse are unfounded. We
could see the unknown light and the lighthouse simultaneously. The latter
was 35 to 40-degrees off where all of this was happening.”[16]

Could Halt and his men have seen the lighthouse beacon and the light
from the object simultaneously? To answer that, the lighthouse keeper at
Orford Ness said this was impossible because a metal bar across the lighthouse lens, in place in December 1980, would have kept the beam from
sweeping the forest floor. Also, he suggested, even if it weren't for the metal
barrier, service personnel from RAF Bentwaters would have recognized the
lighthouse beacon because they would have regularly seen it. In this case,
the appearance of the light was completely anomalous and didn't resemble
the lighthouse beacon in the least.

The Halt Report

Two weeks after the incident in Rendlesham Forest, Charles Halt submitted
a full report to the United States Air Force. The report, written in the last
weeks of the Carter presidency, which Colonel Halt affirmed for *UFO Hunters*, set forth his full description of the incident. It read:

> Department of the Air Force
> Headquarter 81st Combat Support Group (USAFE)
> APO NEW YORK DY735
>
> 13 Jan 81
> Reply to Attn of: CD
> Subject: Unexplained Lights
>
> To: RAF/CC
>
> 1. Early in the morning of 27 Dec 80 (approximately 0300L), two USAF
> security police patrolmen saw unusual lights outside the back gate at RAF
> Woodbridge. Thinking an aircraft might have crashed or been forced

down, they called for permission to go outside the gate to investigate. The on-duty flight chief responded and allowed three patrolmen to proceed on foot. The individuals reported seeing a strange glowing object in the forest. The Object was described as being metallic in appearance and triangular in shape, approximately two and three meters across the base and approximately two meters high. It illuminated the entire forest with a white light. The object itself had a pulsing red light on top and a bank(s) of blue lights underneath. The object was hovering or on legs. As the patrolmen approached the object, it maneuvered though the trees and disappeared. At this time the animals on a nearby farm went into frenzy. The object was briefly sighted approximately an hour later near the back gate.

2. The next day, three depressions 1 1/2" deep and 7" in diameter were found where the object had been sighted on the ground. The following night (29 Dec 80) the area was checked for radiation. Beta/gamma readings of 0.1 milliroentgens were recorded with peak readings in the three depressions and near the center of the triangle formed by the depressions. A nearby tree had moderate (.05—.07) readings on the side of the tree toward the depressions.

3. Later in the night a red sun-like light was seen though the trees. It moved about and pulsed. At one point it appeared to throw off glowing particles and then broke into five separate white objects and then disappeared. Immediately thereafter, three star-like objects were noticed in the sky, two objects to the north and one to the south, all of which were about 10 degrees off the horizon. The objects moved rapidly in sharp angular movements and displayed red, green and blue lights. The objects to the north appeared to be elliptical through an 8-12 power lens. They then turned to full circles. The objects to the north remained in the sky for an hour or more. The object to the south was visible for two or three hours and beamed down a stream of light from time to time. Numerous individuals, including the undersigned, witnessed the activities in paragraphs 2 and 3.

Charles I. Halt, Lt. Col. USAF
Deputy Base Commander

If Charles Halt had hoped that his very inconvenient memo would stir those in both the USAF and RAF or Ministry of Defence higher command to investigate this entire incident, he was to be disappointed. The immediate result of the Halt report was complete silence. His own commanding officer believed that he had overreached by writing the memo and thought that the events the security team encountered were completely conventional. Skeptics and debunkers attacked the witness stories about unidentified flying objects landing in the forest, and the air force itself, according to witness Airman Larry Warren, was subjected to hostile and aggressive debriefings.

Former UK Ministry of Defence spokesperson Nick Pope, himself a respected author and UFO researcher,[17] has said that there were a number of reasons why the British MoD shied away from getting involved in the case. First, he said, higher-ups at the Ministry believed that because American servicemen were involved at what was essentially an American-staffed NATO base, it was the US Department of Defense that was the only entity that should have investigated the incident.

For their part, the Americans seemed to argue, because the actual event took place on UK territory—Rendlesham Forest—and not on the RAF Bentwaters base itself, authority to investigate belonged to the British, thus removing the burden from the US defense establishment to deal with Colonel Halt's report. This was convenient, Nick Pope said on *UFO Hunters* and in a number of radio interviews, because it allowed each side to push the burden of the investigation onto the other side. Accordingly, neither side actually investigated the incident, and the personnel who witnessed the events are still left wondering what really was behind the whole thing.

"You Can't Tell the People"

Pope also pointed to two other interesting post-incident developments as independent researchers in the UK pursued their inquiries. First, British journalist and UFO researcher Georgina Bruni, in her book *You Can't Tell the People*,[18] reported her brief conversation with UK Prime Minister Margaret Thatcher in which she asked the then prime minister about the 1980 RAF Bentwaters/Rendlesham Forest incident. The prime minister was said to have to have replied that when it came to UFOs, "you had to get your facts straight." And that there are some things, presumably relating to UFOs, that "you just can't tell the people." This quote, for many UFO researchers, is one of the most telling statements imaginable because, without admitting to anything, Prime Minister Thatcher admitted to everything. Were there no secrets regarding UFOs, indeed, were there no UFOs at all, the prime minister would have simply dismissed the question or laughed it off. However, because, those of us in the UFO research field believe, the entire subject of the reality of UFOs and what their existence implies is so secret— Margaret Thatcher not one given to outright lying—she simply admitted that the subject can't be talked about at her level because the people, presumably the rest of us, can't be allowed to know the truth.

Even today, the fundamental truth of what happened in Rendlesham Forest has still not been established. For Charles Halt, James Penniston, Larry Warren, and others, they, to this day, assert that they know what they saw. Their case is bolstered by the four prongs of UFO evidence: multiple credible witnesses, documentary substantiation, physical trace evidence, and bizarre government or quasi-government reaction to the event.

Not only do we have the report filed by Colonel Halt, but we have documentation in the form of the "Halt Tape," the voice recording that Charles Halt made on his Lanier device while the actual incident in the forest was unfolding. This ongoing commentary, which was played on History Channel's *UFO Files: Britain's Roswell* as well as on *UFO Hunters,* is probably some of the best documentary evidence available because, to this day, it is the only moment-by-moment description by a highly credible witness in a classified military facility of an anomalous event impacting the security of that facility. Remember, Charles Halt did not go out into the forest to find a UFO. He went out into the forest on a security detail to, in his words, "put a stop" to all the nonsense about a light in the forest. Accordingly, he assembled a highly qualified security detail complete with powerful "Light-Alls," intense lights and secure radio transceivers. What he discovered was not at all what he expected.

The other piece of documentary evidence, in addition to the Halt report, was the official Ministry of Defence corroboration of the high radiation levels in the three landing impressions made by the craft. Because the MoD acknowledged the high radiation levels at the landing site, it, and other documents in the Rendlesham file, amount to nothing less than an "audit trail," as Nick Pope described it, of a UFO encounter. It is all documented and the evidence memorialized in official memos and reports setting forth one of the most complete paper trails of a UFO encounter in any official archive.

Thus, for a US president who tried, but was thwarted in his attempt to reveal the truth to the American people about UFOs, the Rendlesham case, like the Hillsdale case and President Ford in the previous administration, were both tastes of harsh reality concerning the presence of UFOs in the skies over our communities and the official government attempts to cover them up.

Operation Eagle Claw, Cash-Landrum, and Project 160

A defining moment during the Carter presidency, and perhaps the very reason he lost the 1980 presidential election to Governor Ronald Reagan, was the hostage-taking of American diplomats and State Department employees at the American embassy in Tehran in 1979, where Carter, after informally approving of the return of the Ayatollah Khomeini to Iran during a popular uprising there, agreed to bring the deposed Shah to the United States for cancer treatments. This angered the new revolutionary Iranian government, riots in Tehran ensued, the American embassy compound in Tehran was breached, and the Americans inside were taken hostage and kept within the embassy compound. Negotiations failed, and the Carter administration planned a rescue operation code named "Eagle Claw."

The operation was a dismal failure due to a number of incidents: helicopters meant to ferry the American commandos from a rendezvous point in the Iranian desert to the edge of Tehran experienced technical problems including inability to land in the soft powder sand after a sandstorm had covered the landing site, communications problems among the different military services involved in the operation, and the loss of two helicopters, which meant the commander of the operation ordered its termination even before the team made its move toward Tehran.

The failure of Operation Eagle Claw was yet another deep problem for the Carter administration. Sure, military operations fail, didn't one of the aircraft fail during the Navy SEAL raid on Osama bin Laden's compound and didn't the US military commando unit suffer an aircraft loss in the recent raid in Yemen? But this multiple failure of Operation Eagle Claw was looked at by the public as something that happens when a president is unprepared for a serious military confrontation. The planning and execution failures were laid directly at the threshold of the Oval Office. And there was another ironic circumstance to this event when a young US Navy officer who was the navigator on board the destroyer, the USS *Paul F. Foster*, ordered to escort the aircraft carrier USS *Nimitz* into the Gulf of Oman where the *Nimitz*, the carrier from which the mission helicopter would depart, would be on station to receive the hostages and the team that rescued them. But it was not to be.

After the USS *Foster* was ordered to break off their escort of the *Nimitz* and sail to Pearl Harbor, the crew on the *Foster* heard the news that the rescue mission in Tehran had failed with the crash of two helicopters and the deaths of eight military crew members. The officers onboard felt defeated as result of the failure of the mission. One of those officers was a young Stephen Bannon, who would go on after he left the navy to work at Goldman Sachs, to broker the very lucrative deal to take *Seinfeld* into syndication, to become a major player in what was to become known as the "alt-right" in the early news bulletin board groups at the dawn of the Internet, and thence to become the publisher of Breitbart, the brainchild of the late radical right wing conservative Andrew Breitbart, and, ultimately to become Donald Trump's 2016 presidential campaign senior advisor, and President Trump's senior advisor and political strategist at the White House. Quite a journey.

Bannon himself has revealed that he saw the failure of this mission as a much greater failure not only of the military, but Carter's failure as well. Calling his experience as a part of Operation Eagle Claw as "life searing," he said that although he was not interested in politics before the mission, the failure of that mission changed him. He saw the United States failing in the world's political military arena because it couldn't even rescue its own hostages and lost eight of its best military commandos as a result.[19] This event would not only define Bannon's perspective about America's failures, but imbue in him an apocalyptic vision of what the United States must do, even in defiance of its own institutions, to achieve victory in what he saw to be a coming clash of civilizations.

OK, now that we've linked the former White House political strategist and senior advisor, who once sat on the National Security Council, with a failed mission in the Persian Gulf under a Democratic president, where are the UFOs and how many degrees of separation might he have from UFOs?

Answer, probably two. Less than a year after the failed Operation Eagle Claw, President Carter's plans for another rescue mission ran straight into one of the biggest domestic UFO stories of 1980, the Cash-Landrum incident outside of Houston, Texas.

On December 29, 1980, at about the same time that the story of RAF Bentwaters was breaking, Betty Cash, Vicki Landrum, and Vicki's seven-year-old nephew Colby Landrum were driving along a dark and lonely road south of Houston on their way to Dayton, Texas, when ahead of them above the road they spotted a very bright, diamond-shaped object, which seemed to be floating south towards the Gulf of Mexico. Surrounding what they could now make out as a flaming object, the three people in the car could see a small squadron of what had to be double-rotor military helicopters. They seemed to be escorting the object on its way.

As the object came closer, the light from the object became blinding and the heat inside Betty Cash's car was getting intolerable. Betty, who was driving, shut off the air conditioner, stopped the car, and the three passengers got out to watch the object cross the road. But as it came closer, it was so hot, and flames like flashes of lightning were shooting down from the base of the object with such an intensity that Vicki and Colby climbed back into the car, in the back seat, and watched while Betty stayed outside of the car until the heat became too intense, and she tried to get back into the front seat. But the door handle was so hot, it was burning her hand. She tugged on the door harder and harder until it opened and she climbed back into the driver's seat.

Betty, Vicki, and Colby all suffered burns and became sick as they drove home. But Betty's symptoms were worse than the other two. Her eyes puffed up and ultimately closed. Her skin broke out in severe burns. Her hair fell out, and she lapsed in and out of a delirium. Finally, when she did not improve, she was taken to a hospital where she was diagnosed, as were Vicki and Colby, as suffering from radiation poisoning. But how? Her symptoms were similar to those of the survivors on the outskirts of Hiroshima, one doctor said, but there was no nuclear explosion that night and none of the occupants of the car had come into any contact with radioactive material. Or so they thought.

Ultimately, Betty Cash succumbed to her radiation poisoning and the Landrums sued the federal government for their injuries, a suit in tort claiming injuries inflicted upon the plaintiffs by negligence on the part of the government, that was eventually dismissed because the army, a government actor, did not admit culpability in causing any injuries to Betty, Vicki, or Colby. Nor could the plaintiffs, after exhaustive discovery, satisfy the burden of proof that the government caused their injuries. Under any tort claim statute, particularly the Federal Tort Claims Act, the plaintiff has the burden of proving that the defendant was responsible for the damages the plaintiff suffered. Because there was no proof that the government was responsible for plaintiff Landrum's injuries, the suit was dismissed.

Almost thirty years after the incident, a clue emerged during the filming of the "Alien Fallout" episode on *UFO Hunters* when the then head of the Texas National Air Guard suggested that investigators look for any information about what he said was "Project 160," the "Night Stalkers," a very interesting special operations unit with a history dating back to 1980 when Jimmy Carter was planning his second attempt to rescue the hostages held by the Iranians in Tehran. When his first operation, Operation Eagle Claw, failed, the military formed an elite special operations airborne command called Project 160 run out of the Joint Special Operations Command, or JSOC. This top secret unit might have also been responsible for the Cash-Landrum incident, in which this same unit was transporting a flaming radioactive object across a lonely Texas road outside of Houston where it irradiated Betty Cash, Vicki Landrum, and young Colby Landrum.

Because of the planning for the second Iranian hostage rescue attempt, it would have made perfect sense for the military to have kept this Night Stalker unit top secret. However, when the unit was dispatched on an emergency basis to dispose of the flaming nuclear reactor, as they flew it in convoy, they had no warning that they would overfly a car that would stop right beneath it and receive an unhealthy dose of radiation. When Cash and Landrum filed suit against the army for physical damages resulting from the object the Night Stalkers were transporting, the army refused to acknowledge responsibility because they would have had to disclose the existence of the unit and the case against the army was dismissed. The entire story can be seen on YouTube.[20]

Yet here is another UFO-related story during the tenure of President Carter, occurring in the very same month as the story of RAF Bentwaters and another UFO-related incident involving the death of one of the members of the Beatles, John Lennon, and his killer who was carrying a copy of J. D. Salinger's *Catcher in the Rye*. Why?

The UFO Twist to the Murder of John Lennon

Shortly before 11 p.m. on December 8, 1980, as former Beatle John Lennon was exiting his limousine to enter his apartment building, the Dakota, on Manhattan's Upper West Side, a young man named Mark Chapman stepped out of the shadows, called out Lennon's name—although this is in dispute—dropped into what looked like a professional shooter's position—Chapman told an interviewer that's what he did—and fired five rounds into John Lennon's back. This was the same man who earlier that afternoon had asked Lennon to autograph the cover of a record album, *Double Fantasy*.

While Lennon staggered into the building, bystanders knocked Chapman's gun out of his hand, and he was forced to await the police, who arrested him and took him away. Announcements were broadcast over television stations with sportscaster Howard Cosell announcing Lennon's murder. Theories have abounded about why Chapman shot Lennon, something he readily admitted to the police even though he said he could not remember why. Among the theories that circulated was a UFO-themed explanation that Chapman, who came from Hawaii, had been programmed to kill Lennon by the US secret government because Lennon had in his possession a communications device in the shape of an egg that he was said to be using to communicate with extraterrestrials after he had been contacted by them. The story has since been republished in many newspapers and is based on a story from a book by May Pang, who was with Lennon at the time of the incident.[21]

As repeated in the London *Telegraph* by psychic Uri Geller, "They came in the darkness and had bug-like faces. Stranger still, they left a weird egg-shaped object behind."[22] The story that ETs left an egg-shaped communications device with Lennon was reminiscent of a four-hundred-year-old story about an egg-shaped communications device left by ETs with Elizabethan alchemist, astrologer, and England's chief spy, John Dee, reputed to have conjured up the storm that destroyed the Spanish Armada and portrayed as Prospero in Shakespeare's *Tempest*.

Uri Geller told the story about the time he was at a restaurant in New York with Lennon and Yoko Ono when Yoko was soon to give birth to the Lennons' child, Sean. Although John was excited, Uri said, and happy that the couple was expecting a child, John was actually exercised about UFOs. He said that he believed aliens lived on other planets and that they were probably watching people on earth even as they went through their daily lives. Then John told this story, according to Uri Geller: "'About six months ago, I was asleep in my bed, with Yoko, at home, in the Dakota Building. And suddenly, I wasn't asleep. Because there was this blazing light round the door. It was shining through the cracks and the keyhole, like someone was out there with searchlights, or the apartment was on fire.

"'That was what I thought—intruders, or fire. I leapt out of bed, and Yoko wasn't awake at all, she was lying there like a stone, and I pulled open the door. There were these four people out there."

"'Fans?' I [Uri] asked him.

"'Well they didn't want my f—-in' autograph. They were, like, little. Bug-like. Big bug eyes and little bug mouths and they were scuttling at me like roaches.'"[23]

Lennon continued that he wasn't hallucinating or on any drugs but that he really saw little bug-like creatures. That was all he remembered until he awoke next to Yoko who asked him if anything were wrong. In his hands, Lennon was holding a tiny egg-shaped object with a smooth surface. All John knew was that he didn't have the egg when he went to bed, yet he had it in his hands when he woke up. Uri, who still has the object, said that when Lennon handed it to him, he could only see a smooth surface without any graphics or lettering on it. It seemed to have no purpose, unless Lennon himself knew something he didn't know he knew.

The egg is still a mystery, but some UFO theorists have argued that Chapman was a programmed assassin under the control of a black ops group tracking John Lennon because he had been in contact with aliens and was talking about it. Too much truth-telling. Chapman was therefore programmed to kill Lennon, take the blame for it, but erase all memories of the crime from his mind.

A great conspiracy theory, except there's one small overlooked detail in that. Why would Chapman be carrying around a copy of J. D. Salinger's *Catcher in the Rye*? UFO secret-squelching rumors of a programmed killer sent into the world to assassinate John Lennon notwithstanding, the real story has more to do with one of the main points in *Catcher* and in Salinger's life that Chapman, enthralled with Salinger's character Holden Caulfield, projected onto himself and then onto John Lennon.

The actual story that resulted in John Lennon's murder begins with the first publication of Salinger's poignant and painful short story "Uncle Wiggly in Connecticut" in the *New Yorker* magazine,[24] a story about two former college roommates, Mary Jane and Eloise, and Eloise's daughter Ramona and Ramona's imaginary friends Jimmy and Mickey, her husband, Lew—with whom she is drowning in the backwash of the debris of marriage—and her first serious boyfriend, Walt—killed in a ridiculous explosion of a Japanese field stove in the South Pacific during World War II—who could make her laugh without trying to be funny. And now, here she is, sitting across from her friend, career-girl Mary Jane, explaining why she still missed Walt after all these years now that she was trapped in a loveless but convenient suburban marriage to Lew and dealing with a daughter for whom she feels only an unraveling thread of empathy.

Eloise tells Mary Jane a story that typifies the relationship she had with Walt by describing an incident when she had fallen outside the PX while waiting for Walt, who was late, and twisted her ankle as the two of them ran after the bus that was pulling out. "Poor Uncle Wiggily," Walt said, soothing Eloise, "Poor old Uncle Wiggily." Eloise remembered. "God, he was nice."[25] "Uncle Wiggily," originally Uncle Wiggily Longears, a rheumatoid arthritic rabbit who walked with a cane, was one of the characters in a series of illustrated children's books by Howard Garis originally writing for the *Newark News*; here, he becomes the objective correlative of Walt's humor and kindness for a hobbled Eloise, whom he comforted. That comforting image of Uncle Wiggily is what Eloise clings to, years later, to assuage her display of self-hatred and fury when, after her daughter Ramona tells her that her imaginary friend, Jimmy Jimmereeno, has been run over by a car and killed. Later in the story, when a now thoroughly inebriated Eloise finds Ramona sleeping at the edge of her bed so as not to crush her new replacement imaginary friend, Mickey Mickereeno, Eloise forces the crying Ramona to sleep in the center of the bed, and then, in a rush of self-pity at love lost, presses Ramona's glasses to her cheek and laments, "Poor Uncle Wiggily." Why should Ramona enjoy the fruits of an imaginary relationship while Walt, who still lingers in Eloise's fantasy of love, is extinguished in real life? And, thus, Eloise forces Ramona to crush Mickey Mickereeno, extinguishing him as Eloise has had to extinguish Walt.

As poignantly melancholy as this story is—true love wiped out by war, a convenient marriage for which a child must suffer, a young woman who still laments the one person who brought her comfort, and a moment of memory shared with an old college roommate in a sterile suburban home on a slushy wintry day when memories of a time before the innocence of childhood had given way to the harshness of adult life—Eloise's admission to herself that she has prostituted herself to a lifestyle that is not her, has become a phony, precedes her pleading with Mary Jane to tell her that once upon a time, Eloise was a "nice girl."

But where's the link to John Lennon?

In 1949, a year after the publication of "Uncle Wiggily in Connecticut," Samuel Goldwyn's film company purchased the story for a motion picture, retitled *My Foolish Heart* (remember the title song?), starring Dana Andrews and Susan Hayward as Eloise, but this time throwing in a twist. Lew, now married to Eloise, was once Mary Jane's boyfriend, and the poignancy of Salinger's setting in juxtaposition the four aspects of love and remembrance was simply gone. J. D. Salinger, himself, was furious and believed that he had been artistically deceived by Hollywood.[26] Thus, in 1951 when Salinger published *Catcher in the Rye*,[27] one of the most important novels of its generation, the main character Holden Caulfield rails against his brother, D.B., who prostituted himself in Hollywood and became a phony. His younger brother Allie had died of leukemia while still in the innocence of childhood while his younger sister, Phoebe, is on the very edge at the end of childhood, but still innocent. Holden, himself confronting the harshness of the adult world in New York City, vows that he will be the protector of childhood innocence, catching the innocent as they walk through the rye fields of life, unaware of the obscured dangerous cliff that lies before them. Holden will keep them from falling over that cliff by catching them before they fall.

And, thus, Mark Chapman, perhaps seeing that John Lennon, whose songs about innocence and freedom, were of a fleeting moment, and fearing that his idol of the age of the Beatles years earlier was losing his innocence, was becoming one of "them," realized that Lennon had to be protected lest he, too, not see the cliff though the rye and fall off. Mark Chapman did exactly what he believed Salinger, in the voice of Holden, was telling him to do: protect Lennon. And in his delusion he did it in the way he believed. In this way, Mark David Chapman *became* the Catcher in the Rye.

"If you want to know the truth."

Chapter 15

PRESIDENT RONALD REAGAN AND UFOS

According to first daughter Patti Davis, her father Ronald Reagan was fascinated, almost transfixed, by the subject of UFOs. Like his predecessor Jimmy Carter, Reagan had personally seen UFOs on at least two occasions and, more than likely, he had been briefed on the subject of UFOs when he first assumed the presidency. At least his vice president, George H. W. Bush, was in the know about UFOs even if only because of his former position as the Director of Central Intelligence. We believe that Reagan's sightings took place once in the 1950s and again in 1974 when he was the governor of California, a multiply witnessed sighting.

Mojave Desert, 1974

"I was in a plane last week," California Governor Ronald Reagan recalled to *Wall Street Journal* reporter Walter Miller, "when I looked out the window and saw this white light. It was zigzagging around. I went up to the pilot and said, 'Have you ever seen anything like that?' He was shocked and said, 'Nope.' And I said to him, 'let's follow it!'"[1]

"We followed it for several minutes. It was a bright white light. We followed it to Bakersfield, and all of a sudden to our utter amazement it went straight up into the heavens. When I got off the plane I told Nancy about it," Governor Reagan affirmed to Miller before he realized, registering the shocked look on Miller's face, that, exuberance notwithstanding, the governor was disclosing to a member of the press his having seen a UFO.[2]

Governor Reagan was flying back to Sacramento when one of the three passengers on his Cessna Citation pointed out to the pilot, Bill Paynter, an object he could not identify that was flying a couple of football fields away behind the plane. Reagan asked that the pilot follow the object, which had started to accelerate after the pilot banked the plane to get a better view of it. "It was a fairly steady light until it began to accelerate. Then it appeared to elongate. Then the light took off. It went up at a 45-degree angle-at a high rate of speed. Everyone on the plane was surprised. . . . The UFO went from a normal cruise speed to a fantastic speed instantly. If you give an airplane power, it will accelerate—but not like a hot rod, and that's what this was like," pilot Bill Paynter remembered.[3] And before anyone could even get a closer description of the craft, it climbed out of sight over California's Mojave Desert and was gone in a flash. The governor was very excited about what they saw and within the week, he had told Walter Miller all about it, adding that he was also telling Nancy about it.

The Pacific Coast Highway Sighting

Actress Shirley MacLaine, an author in her own right, related a story told to her by fellow actress, television star, and comedian Lucille Ball about an event with Ronald Reagan back in the 1950s while he was still an actor. The Reagans were driving along PCH, the scenic Pacific Coast Highway that runs along the California coast, on their way to a party for actor William Holden when, according to Lucy, as reported by MacLaine in the UK's *Daily Mail*, "that a UFO landed and the alien emerged telling Reagan to quit acting and take up politics."[4] In the 1950s, Ronald Reagan had migrated to television and was promoted by the Hollywood and New York advertising agencies as a perfect corporate spokesman. It would have been imprudent for him, at the least, to start talking about UFOs to anyone other than those in his LA entertainment community. Thus, his excitement relayed to his friend Lucille Ball, also a major hit in the 1950s, along with husband Desi Arnaz who started Desilu Studios, the initial producer of Gene Roddenberry's *Star Trek*.

Reagan's 1981 CIA Briefing

As recalcitrant as George H. W. Bush might have been about debriefing President-elect Jimmy Carter concerning any government UFO files, that hesitation disappeared when the former DCI became Ronald Reagan's vice president. Bush's successor at the CIA was William Casey, a former OSS officer who had served in Europe during the war. Casey, along with the OSS Director William "Wild Bill" Donovan, FBI Director J. Edgar Hoover, and former New York prosecutor and later governor Thomas E. Dewey, instituted one of America's most successful wartime counterintelligence programs known as "Operation Underworld." This was an operation in which former mob boss Charles "Lucky" Luciano was moved from prison in upstate New York to Sing Sing, just north of New York City, so that he could confab with mob leaders like Frank Costello and union leader Joe "Socks" Lanza to ferret out Mussolini's spies on the New York waterfront, who were tipping off merchant shipping information to Nazi U-boats lurking offshore. The operation was so successful that some of the mob-connected soldiers who worked for this operation on the New York waterfront were later recruited by Hoover to form an ad hoc group of black bag contract killers to eliminate spies that had managed to remain in the US after the war.[5] But in early 1981, Casey was the Director of Central Intelligence for Ronald Reagan and at a cabinet meeting in February of that year made the prophetic statement that, perhaps inspired the folks in today's alt-right demographic, perhaps even Breitbart's Steve Bannon, when he said, "We'll know our disinformation program is complete when everything the American public believes is false," which is perhaps all we actually need to know.[6]

Bill Casey was also one of the special presidential briefers the following month at Camp David when President Reagan along with Secretary of Defense Caspar Weinberger and Reagan's Deputy Chief of Staff Michael Deaver were updated on what the intelligence community knew about UFOs, a subject that excited President Reagan. According to what we believe are notes and transcripts from that briefing, the conversation began with Casey's stressing the briefing itself would be "sensational," detailing chronologically the presence of an extraterrestrial intelligence operating in Earth's skies and on our planet. Referring to the content of the briefing as "Above Top Secret," Casey explained that the general overview of the information that was about to be revealed had been shared months earlier, and Reagan said that he had come into contact with some of the government's information back in 1970 when Richard Nixon had shared some of the files with top colleagues in the Republican Party. This was a stunning statement because, if the transcript is true, it means that Nixon not only shared what he knew with Jackie Gleason, he circulated the information to a larger group. According to President Reagan, Nixon was "fascinated by it."[7]

Referring to President Nixon, Ronald Reagan revealed, "He showed me something, some kind of object or device that came from one of their craft.

Something that was taken from the New Mexico crash site. I don't know if, well . . . huh . . . do we know what it was?"[8]

After pleasantries, a short break, and a quick discussion about who from the White House and security staff should remain in the briefing room, one of the briefers, unnamed in the transcript, begins with a general introduction of how the information about UFOs first got into government files: "The United States of America has been visited by Extraterrestrial Visitors since 1947. We have proof of that. However, we also have some proof that Earth has been visited for many thousands of years by various races of extraterrestrial visitors. Mr. President, I'll just refer to those visits as ETs. In July 1947, a remarkable event occurred in New Mexico. During a storm, two ET spacecraft crashed. One crashed southwest of Corona, New Mexico and one crashed near Datil, New Mexico. The U.S. Army eventually found both sites and recovered all of the debris and one live Alien."[9] The briefer also refers to the individual extraterrestrial inhabitants of the craft that crashed outside Roswell as EBEs, standing for extra biological entities, a term used in the Eisenhower MJ-12 briefing documents as well. The briefer revealed that the debris from the New Mexico crash sites, including the bodies of aliens from the spacecraft, eventually wound up at Wright Field in Dayton, Ohio, later renamed Wright-Patterson.

Asking for a description of the EBEs, Reagan was told, "They don't have any similar characteristics of a human, with exception of their eyes, ears and a mouth. Their internal body organs are different. Their skin is different, their eyes, ears and even breathing is different. Their blood wasn't red and their brain was entirely different from human. We could not classify any part of the Aliens with humans. They had blood and skin, although considerably different than human skin. Their eyes had two different eyelids. Probably because their home planet was very bright."[10] Then the briefer explained where these EBEs were thought to have come from, and the briefer explained, seeming to confirm what Betty Hill said her ET abductor told her, "A visitor from another planet . . . another world. EBE did explain where he lives in the universe. We call this star system Zeta Reticuli, which is about 40 light-years [38.42] from Earth. EBE's planet was within this star system."[11]

When the president asked about the format of the briefing, whether this information was to be revealed in stages or whether there would be a full data dump, it was explained to him by other briefers in the room that the government had been engaged in a full disinformation program since 1947 and for the purposes of understanding the extent of what the government knew, it was better for the briefing to proceed slowly and by stages, and with some information withheld so that the president would have plausible deniability with respect to facts about UFOs and ETs. As Bill Casey put it to the president, he could ask any questions he wanted because "You are the President. We are not here to argue with you over the order of this briefing. But some things are so highly classified that this briefing is the lowest level. If you ask a question that is in a different level, then we will have to re-evaluate the audience."[12]

The aliens traveled to earth from the Zeta Reticuli system, the briefer explained, "It took the EBE spaceship nine (9) of our months to travel the 40 [38.42] light-years. Now, as you can see, that would mean the EBE spaceship traveled faster than the speed of light. But this is where it gets really technical. Their spaceships can travel through a form of 'space tunnels' that gets them from point A to point B faster without having to travel at the speed of light. I cannot fully understand how they travel, but we have many top scientists who can understand their concept."[13]

The briefer explained that although the records of UFOs were maintained under different file names since 1947, briefing papers under the Truman administration were provided to the incoming President Eisenhower and a panel of experts was formed known as MJ-12, and ultimately Project Blue Book was established to provide a public face to the general public interest in UFOs; that project was later abandoned. The briefer said, "MJ-12 decided that officially the Air Force should end their investigation of UFO sightings. This decision was arrived at during the NPNN meeting in 1966. The reason was twofold. First, the United States had established communications with the aliens. Second, the government had to continue the cover-up of the entire UFO question."[14]

Talk about a showstopper! But hold your horses, because in the conversational exchange memorialized in the transcript of the briefing, President Reagan refers to Weinberger as "Casp." According to those who knew both men and who were at other briefings with the president and the secretary of defense, Weinberger was only called "Cap" by his friends and by the president, not "Casp." Perhaps this severely impairs the credibility of this transcript. Or maybe it's just a typo, someone hitting the "a" and "s" keys at the same time.

Nevertheless, moving on, before President Reagan could get the briefer to explain what he meant about communication with the aliens, the briefer explained that the research into ETs suggested that they were no threat to planet Earth or to human beings. However, the government realized that there was a level of existential threat to human institutions from the realization that other intelligent life forms inhabited the universe and had arrived on Earth. Because the government and church leaders believed that the alien presence was a threat to religious institutions and the resulting panic from an ET presence might destabilize societies, the entire study of UFOs was covered up. And this cover-up remained in place even after a level of communication was established between governments and ETs. However, the intelligence community sought to "normalize" human acceptance of ETs and UFOs by funneling stories about them into the entertainment community as a form of entertainment/disinformation. As the briefer explained, "In order to protect all this information and the fact that the United States Government has evidence of our planet being visited by extraterrestrials, we developed over the years a very effective program to safeguard the information. We call it 'Project DOVE.' It is a complex series of operations by our military intelligence agencies to disinform the public." In fact, the briefer explained, "The first cooperative venture was the movie, *The Day the Earth Stood Still*. That was a cooperative venture with the United States Air Force and the movie industry."

This prompted President Reagan to ask, "That movie, *Close Encounters*, was that one of them? I guess no."

"Yes," the briefer answered, and then went on, ""Yes, Mr. President, we provided the basic subject matter for that movie."[5] Wait, just one more thing, did anybody tell Steven Spielberg?

President Reagan and Steven Spielberg

Which now brings up the question: what, if anything, did President Reagan and Steven Spielberg talk about with respect to extraterrestrials after the special screening of Spielberg's *E.T.* at the White House? Rumors about the conversation President Reagan had with Steven Spielberg were rampant, with no confirmation from any of the principals, which suggested that with a presidential wink and a nod both men agreed that the movie *E.T.* was certainly more science than fiction. In an interview with "Ain't It Cool News," Steven Spielberg said during post-screening interview for *Super 8* concerning the rumor that, during a screening of *ET*, President Reagan was ushered out of the room before speaking to him:

"No, he wasn't ushered out of the room. He was the President of the United States! Nobody could usher Ronald Reagan out of the room! It was in the White House screening room and Reagan got up to thank me for bringing the film to show the president, the First Lady, and all of their guests, which included Sandra Day O'Connor in her first week of as a Justice of the Supreme Court, and it included some astronauts . . . I think Neil Armstrong was there, I'm not 100% certain, but it was an amazing, amazing evening.

"He just stood up and he looked around the room, almost like he was doing a headcount, and he said, 'I wanted to thank you for bringing *E.T.* to the White House. We really enjoyed your movie,' and then he looked around the room and said, 'And there are a number of people in this room who know that everything on that screen is absolutely true.'

"And he said it without smiling! But he said that, and everybody laughed, by the way. The whole room laughed because he presented it like a joke, but he wasn't smiling as he said it.

"The room did laugh and then later on I'll never forget my conversation with the president. He pulled me aside, he said . . . and I can't do Reagan. I wish I could do that breathy, wonderful voice of his . . . And Nancy Reagan was standing right next to him and the President said to me, 'I only have one criticism about your movie,' and I said What's that? He said, 'How long were the end credits?' I said, Oh, I don't know. Maybe three, three and a half minutes? He said, 'In my day, when I was an actor, our end credits were maybe fifteen seconds long.'"

"He said, 'Why don't you let everybody get a credit . . . three and a half, four minutes, that's fine, but only show that inside the industry, but throughout the rest of the country reduce your credits to fifteen seconds at the end?' Nancy Reagan turned to him and said, "Oh, Ronnie, they can't do that. You know that.' And he went, 'Oh, yes, yes. I suppose.' (laughs) That was the extent of my conversation about that. That was his only criticism, he felt the end credits were too long!"[17]

After Truman, after JFK, after Carter, and now after Reagan, how much more disclosure will prove the point?

President Reagan's Alien Invaders Speech

"In our obsession with antagonisms of the moment," President Reagan said to the United Nations General Assembly on September 21, 1987, "we often forget how much unites all the members of humanity. Perhaps we need some outside, universal threat to make us recognize this common bond. I occasionally think how quickly our differences worldwide would vanish if we were facing an alien threat from outside this world."[18]

The following year, after his meeting in Reykjavik, Iceland, with Soviet leader Mikhail Gorbachev, "I occasionally think how quickly our differences, worldwide, would vanish if we were facing an alien threat from outside this world."[19] As former CBS News anchor Dan Rather might say, "What did he know and when did he know it?" And in that same year, when President Reagan was writing his "disarmament" speech to the UN General Assembly, he again made reference to threats from outer space, specifically writing in the draft speech the phrase that "an alien force is already amongst us."[20] When, however, his speechwriter took his self-described "fantasy" about an "alien force already against us," the president, in his handwritten notes to his speechwriter, stated that he wanted his "alien force" statement back.[21] That, as one can imagine, is a very important, even if backdoor, disclosure.

The Hudson Valley Sightings and Indian Point

Almost a year after Ronald Reagan was allegedly briefed officially on the UFO question, a stunning series of UFO events took place over New York's Hudson Valley about an hour north of New York City and over the nuclear power plant at Indian Point. Shrouded in mystery and a serious attempt at debunking, the Hudson Valley sightings involved stories of sightings of humanoid creatures, abductions, and the participation of Blue Book science analyst and former UFO debunker J. Allen Hynek.

The first sighting, and for a fuller description see the *UFO Casebook*,[22] took place in Kent, New York, when a retired police officer sitting out in his backyard on New Year's Eve 1981 saw a group of colored lights in the sky overhead. Lights in the sky were not, in themselves, an anomaly this far north of the City, not with three different metropolitan airports handling intercontinental and international flights—however, even given all the colors that might have been an airliner's navigation lights, the observer was still confused because something flying that low would have made lots of noise if it were a conventional airplane. This configuration of lights was completely silent and, worse, the witness noticed, the lights were moving too slowly for a conventional aircraft.

As the lights came closer, the witness realized that these weren't just a constellation of separate illuminated objects, they were fixed to a rigid structure, a giant triangle. The witness reported that as the triangle floated directly over the backyard, he could hear a soft hum in the air. He realized that he was looking not at a conventional aircraft at all, but a UFO.[23]

Into 1982 and 1983, the early years of Reagan's presidency, the sightings of triangular flying Vs would continue throughout the Hudson Valley area with scores of witnesses, at times lining the roads like the Taconic State Parkway to watch the light formations transit across the sky. On at least one occasion, a witness saw a giant triangle hover just over the Croton Falls Reservoir and shine a searchlight-type of beam over the water as if it were looking for something or maybe even electrolyzing the surface to pull up specific chemicals in the water.

The Storyville Flyers

Of course, the skeptics poured their derision on the witnesses to the Hudson Valley sightings, claiming that the observers were only looking at aircraft and nothing more. Then, as if to validate the skeptics and debunkers, a group of flyers who called themselves the "Storyville Flyers," pilots who were practicing close formation flights, popped to say it was all a hoax.[24] By 1983, as the stories of the Hudson Valley sightings were capturing the attention of local newspaper and television reports from the Bronx to New Haven, Connecticut, the group of amateur pilots claimed that they were the source of the Hudson Valley sightings, lights in formation. They claimed that when they heard the stories of UFOs over Hudson Valley, they decided to fly in a closer formation and give the witnesses something to shout about. However, UFO researchers investigating their claims said that although some of the pilots' flights might have been mis-taken for Hudson Valley lights, their basic claim that they were the sole cause of the lights made no sense because some of witnesses, including the witness on the Taconic State Parkway, were absolutely convinced that the lights were part of a large, multi-football-field-sized rigid triangle, not a formation of separate planes. Besides, as witness after witness said, the overflights of the lights were completely silent except for an occasional hum when the lights were at about a hundred feet in the air. A forma-tion of propeller-driven planes would have made a thunderous amount of noise flying that slowly and at that altitude. The Hudson Valley lights could not have been the Storyville Flyers. For an interesting video of one of the Hudson Valley lights triangles, go to http://www.abovetopsecret.com/forum/thread1101879/pg1.

The sightings became so prevalent that a local conference was set up for the witnesses to share their experiences, and in attendance was the UFO researcher J. Allen Hynek, the scientist who had worked as an investiga-tor for Blue Book and who coined the famous term "swamp gas" about the nature of the Hillsdale, Michigan, sightings that so exorcised President Ford he demanded Congressional hearings on UFOs. Hynek was not dis-suaded from according credibility to the sightings and even talking to some of the conference attendee/witnesses who had claimed missing time during the lights overflights and abductions. Hynek was no longer a skeptic at this point, he said, having been convinced by his research into Blue Book topics that there was more to UFOs that needed to be investigated. Hynek specifi-cally asked many of the witnesses to gather as much evidence as they could and write complete reports about their sightings. More to the point, accord-ing to more than one guest on the History Channel's *UFO Hunters*, who, for obvious reasons, shall go nameless, he revealed that early on in his associa-tion with Blue Book, Hynek had been taken to see a live ET that was being kept for testing purposes by a group within the Eisenhower administration. Hynek's access to this live ET was premised, our sources told us, upon his being the official skeptic and debunker for Blue Book.

Indian Point

Perhaps the most potentially dramatic and threatening episode during the Hudson Valley UFO flap was the appearance of a giant flying triangle over the Indian Point nuclear power plant north of New York City in 1984. The UFO appeared on two separate occasions over Indian Point, June 14 and June 24, startling guards and workers. The June 14 sighting was only a fly-over, but it still alerted guards because nuclear power plants are in no-fly zones heavily monitored and heavily punished for infractions. This flyover, too, eliminates the Storyville pilots because even a bunch of amateur hoax-ers looking to stir up trouble aren't dumb enough, if they are serious pilots, to put their FAA licenses in jeopardy by flying close to the ground in a no-fly zone.

On June 24, one of the security guards who first spotted the triangle called out, "here comes that UFO again."[25] He alerted the other guards as the craft seemed to hover about 300 feet over the number 3 reactor, the only reactor online that evening. The security guard, who is also a nameless source, estimated that the triangle was over 1000 feet and simply hovered in place. It wasn't a plane or a balloon; it was a giant triangle. The length of time the craft spent over a live reactor, and its size, prompted the security management team to consider calling out the Air National Guard for a flyby look-see at the object. There were debunkers, of course, who said that the object was not a triangle, but actually a blimp, a thousand-foot blimp. However, why would a blimp float over a nuclear power plant, hover there, and flash its lights for no apparent reason but to draw attention to itself and cause a panic? It makes no sense.

In the aftermath of the giant triangle flyover on June 24, one of the researchers of the Hudson Valley sightings said that he learned that the number 3 reactor, the reactor that was live at the time of the June 24 incident, developed a crack in it after the incident was over. While one of the managers at the plant denied any cracks in the reactor, UFO researchers of this event hypothesized that UFOs had been overflying nuclear facilities since 1945 looking to see how we were handling our nuclear enrichment and energy apparatus. Perhaps the UFO determined by scanning that there was a flaw in the reactor wall at number 3, assessed the likelihood of a nuclear meltdown from containment failure, found that that the crack was not structurally critical, and flew off after deciding not to shut the facility down.

For a president like Ronald Reagan, however, the incident itself, com-ing on top of a multi-year UFO flap over New York and after having been briefed about UFOs, including his own sightings, one can imagine that he would have been more than involved in learning about all the details of this incident.

Japan Air Lines Flight 1628

Is the president of the United States briefed or even copied on every UFO sighting incident? Probably not. But is it conceivable that the president would be copied if the military, of which he or she is the commander in chief, had a direct encounter with a UFO? We think that might be the case for the November 17, 1986 encounter of a Japan Air Lines Boeing 747 commercial cargo flight on its way to Japan from Iceland. That night, JAL Flight 1628 encountered a UFO off the coast of Alaska and radioed that contact. The radar contact was picked up at Elmendorf Air Force Base in Alaska, and both the pilot and copilot had a visual sighting of the bright floating light outside the aircraft. The records of that sighting and radar contact were assembled by the FAA. The CIA later walked into a meeting of a discussion of JAL 1628 and removed all of the documentary evidence from the FAA and then handed the it over to their own UFO consultant, Dr. Bruce Maccabee, who then wrote a detailed report on the event.[26]

The object was first spotted, according to Dr. Maccabee's report, just after 5:11 p.m. local time over northeastern Alaska by the JAL pilot Captain Kenju Terauchi, who said the object, a large floating light, was to the left and rear of his aircraft. At first, Terauchi had no concerns about what he saw because he believed there to be a group of American military jets in the vicinity because the JAL flight was near two air force bases and was closing in on the boundary between American and Soviet airspace. But things changed rapidly when, after changing his course heading per instructions from air traffic control, Captain Terauchi realized that he had not left the light behind—it had changed course, too, and was following him. But then, in a way that no conventional aircraft could have done, the configuration of lights suddenly jumped from aft of the portside of the aircraft to the front of the aircraft's cockpit window, which was enough to cause concern to the cockpit crew.

According to Captain Terauchi's own testimony after the event, "It was about seven or so minutes since we began paying attention to the lights [when], most unexpectedly, two spaceships stopped in front of our face, shooting off lights. The inside cockpit shined brightly and I felt warm in the face."[27] What has piqued the interest of UFO historians was Captain Terauchi's reference to the lights as "spaceships" and his description of the lights' seeming use of thrusters, whether jet thrusters or rocket thrusters, to maneuver themselves into position.

As an experienced pilot, Terauchi radioed air traffic control about the lights that were so close the heat from them was entering the cockpit. Rather than "UFO sighting report" per se, the pilot, as was required, was asking the air controllers if they were picking up any air traffic at JAL 1628's twelve o'clock; that there was a collision possibility. ATC said they had no traffic in the area, nor could they pick up anything on radar that posed a danger to flight 1628, but asked the pilot for the position of the lights he saw. Terauchi replied that he believed he was looking at different colored flashing navigation lights that were strobing.

Of greater concern to JAL's cockpit crew was what took place when they tried to take photos of the object in front of the plane. When they focused a camera that was in the cockpit, at first they could not get a proper focus because the autofocus on the camera was not working. Then when Captain Terauchi tried to switch over to manual focus, he could not get the camera's shutter to operate properly. And right after that attempt, the entire plane began to shake as if being buffeted by an outside force, in response to which Captain Terauchi stowed the camera and tried to stabilize the aircraft while at the same time the air traffic controllers contacted Elmendorf Air Force Base to see if they had anything on radar close to Flight 1628.

Suddenly the lights in front of Flight 1628 changed again with what seemed to be a merging of the flashing lights into a larger light that the JAL pilot said looked like a craft of some sort. This time, while Terauchi was certain that his visual sighting was accurate, the radar operators at Elmendorf said they could not pick up returns on their radar screens, nor could they confirm what the JAL flight crew saw. But Terauchi was not satisfied and checked his plane's weather radar to see whether the object he could see with his own eyes turned up on his radar screen. It did. And this time the object showed up on the traffic control ground radar as well, and then it moved from the front of Flight 1628 to the rear, at which point radar at

Elmendorf picked it up. From Terauchi's description of what he could see in coming twilight, the object bore no resemblance to any existing conventional aircraft. It was, in Terauchi's words, a "mothership."[28]

Behind him and against the background of ground illumination, Terauchi could now see the outlines of a large craft, again, unlike any conventional airplane, which prompted Terauchi to request his flight controllers to grant him permission to make a course change of forty-five degrees to inspect the object following him. The object changed its course as well and continued to follow Flight 1628 through a number of course changes for which Terauchi requested permission that was granted. Another commercial airliner was in the area, which the traffic controllers vectored toward Flight 1628 to ascertain if they could get a visual, which they could not even though the object, which was moving away from the JAL flight, could still be seen on radar. Finally, Flight 1628 proceeded to its destination, but the event had such an impact on Captain Terauchi that he decided to report it after he landed and cooperated in an FAA investigation headed up by the FAA's chief investigator, John Callahan, who, over twenty years later, was interviewed on the History Channel's *UFO Hunters*. The entire story of JAL Flight 1628 along with the radio transmissions from the pilots to air traffic control appeared on the History Channel's *UFO Files*.[29]

What Callahan explained to his interviewers was even more intriguing when he stated that at a meeting to review all of the archived radio and radar material, personnel, who identified themselves as intelligence officers, entered the room, and under federal national security authority, took possession of all the material. Fortunately, Callahan explained to the UFO Hunters, he had made duplicates of all the material he was told to hand over. By the way, Callahan also explained to his interviewers, TWA Flight 800 was shot down by a navy missile in 1996.[30]

It would be naive to think that a commercial aircraft encounter with an unidentified object that aggressively followed that aircraft and was picked up on military radar right at the airspace border between the US and the USSR would not have been brought up to President Ronald Reagan. Here was a president, near the end of his term, who had borne personal witness to UFOs on at least two occasions, had been the recipient of other UFO reports, had dared to broach the UFO issue to the Soviets and to the

United Nations, and who fiercely advocated for a planetary defense system he called the "Space Defense Initiative," which, we believe, was intended not only to shoot down enemy ballistic missiles but to observe and possibly interdict UFOs as well.

Chapter 16

GEORGE H.W. BUSH: "AMERICANS CAN'T HANDLE THE TRUTH!"

Bush and Carter

To deconstruct the statement George Bush made when he was Director of the Central Intelligence and told the incoming President Jimmy Carter that he had "no need to know" about the contents of CIA files on UFOs, one should look at how what he said opened a very big door. He was essentially admitting that not only did the United States have UFO files but that they were too sensitive and held at a classification above the president-elect's top secret clearance. This was a stunning admission. But, as documented on Grant Cameron's Presidential UFO website,[1] even more stunning was a public statement Bush made in 1988 to a UFO researcher, Charles Huffer, when he was running for president. He was asked whether the United States would tell the truth to the American people about UFOs.

"Yeah," he replied. But, as the candidate rushed inside to meet reporters, did he recognize the import of the question? The former Director of Central Intelligence, then current vice president and candidate for the presidency, had just acknowledged that there were UFO secrets the government was keeping and that he would release them. Was this what he meant? To qualify what he said as he was hustled away, he said, "If we can find it, what it is. We are really interested."[2] Charles Huffer asked again whether the candidate would declassify any information on UFOs—"Declassify it, OK?"—he could after he became president, and Bush assured him, "OK. Alright, yes."[3] And when exiting the meeting with reporters and again encountering Huffer, who reminded him of his promise and assured him that information was already in the government files, Vice President Bush said, "I know some. I know a fair amount."[4] Sometimes when political leaders are answering questions in a rush, the truth really does come out.

The Brooklyn Bridge Abductions

The entire subject of alien abductions is rife with controversy. However, the Brooklyn Bridge abductions, popularized by researcher Budd Hopkins in his book, *Witnessed*,[5] became a major story in 1990 regarding a stunning scenario of a UFO beaming a young mother though the window of her apartment building in lower Manhattan near the Brooklyn Bridge. The ET motives for abductions of human beings vary from the hybridization of the species, combining human DNA with alien DNA to establish a hybrid human presence on Earth that would replace human beings, to warning human beings about the dangers of a nuclear war that could wipe out life on this planet or about the dangers of climate change that could also render human life on Earth extinct. Pick an alien abduction researcher and pick the theory. The problem with alien abduction research, however, is that some of the techniques of retrieving information supposedly lost during the abduction itself involve forms of therapy that sometimes fly in the face of established medical procedures. Chief among these procedures practiced by self-described abduction therapists is hypnotic regression, wherein the hypnosis itself isn't the immediate problem, but, as Budd's former spouse and filmmaker/author Carol Rainey has explained, the preconditioning prior to hypnosis and the leading of a subject during the hypnosis that affects the way the subject may remember a traumatic event. In prolonged hypnotherapy treatments performed by those not certified clinical psychologists or psychiatrists, in short, not trained medical personnel, the therapist is likely unaware of a transference and counter-transference that takes place in which the subject seeks to please the therapist and the therapist, in turn, seems to frame the therapy to support what the subject is saying regardless of its implausibility.

These questions about the accuracy and efficacy of alien abduction regression therapy lie at the root of many debates over the veracity of stories told by those who claim to be abducted. For example, in the Betty and Barney Hill case, although both Betty and Barney consciously remembered being followed by a light in the sky, stopping on the road by a craft in front of them, and being dragged into a clearing by strange looking humanoid creatures, there was no preconditioning on the part of their therapist Dr. Benjamin Simon, who was as surprised as anybody upon hearing the Hills relate their stories under hypnosis. In more recent stories of abduction, many a researcher talks with the subject in a waking state about the event and in that conversation subtly plants clues in the subject's mind about what he or she is about to say about the event. This, Carol Rainey has said, is what happened in many of Budd's "Intruders" group sessions with subjects before they underwent hypnotic regression therapy with him.

With this as a general caveat regarding the methodology of researching alien abductions, most people in the UFO community try to rely more on extrinsic witness reports of events that might center around a single person's recovered memories. For example, in the Brooklyn Bridge abductions, people came forward who claimed to be on either the Brooklyn or Manhattan Bridge on the night of November 30, 1989, when Linda Napolitano—Budd referred to her as Linda Cortile—said that she had been abducted. According to Hopkins, as reported by researcher Grant Cameron on his Presidential UFO website, one very important witness had communicated with Linda after her claimed abduction, saying that he had witnessed the entire event. Although the veracity of that correspondence has been questioned, the person who wrote the letter was the United Nations Secretary General Javier Perez de Cuellar, who was on the bridge that night on his way into Manhattan and who, Budd Hopkins theorized, wasn't just a witness, he was an actual abductee along with Linda. Perez de Cuellar was traveling, according to what Hopkins learned, with two Secret Service agents. This is an important fact, even if only for the theory that any event regarding an abduction or kidnapping involving Secret Service personnel would have immediately set of red alerts throughout the agency and would have most certainly reached the desk of President George H. W. Bush.

According to Grant Cameron, "After further investigation, Hopkins determined that Perez de Cuellar was not a witness, but had been abducted as well. This fact was confirmed by Linda's son who identified the Secretary General from twenty photos of men shown to him by Hopkins, as the man who was with him and comforted him *during* the abduction. Hopkins contacted Perez de Cuellar in person about the letters he had sent to Linda and himself, and de Cuellar didn't deny his involvement."[6]

The ETs' rationale for this abduction of de Cuellar, supposedly, was to take him to an island in the South Pacific that was going to disappear beneath the ocean to demonstrate that climate change would soon render many parts of the planet uninhabitable and that, as Secretary General of the United Nations, de Cuellar was in the best position to warn nations about this coming climate catastrophe and instigate changes in the way nations dealt with the environment. Years later, de Cuellar denied the entire story.

The Gulf Breeze Sightings

When UFO incidents take place near important military bases. news of the incident, even when it involves civilians, travels up the chain of command and winds up among the Special Access Projects groups for quiet review. When the events stretch over years and involve multiple witnesses and larger groups, more often than not, operatives from the Special Access Groups mobilize resources to debunk the events and disparage the reputations of the witnesses, thus throwing their credibility into question. If you don't believe the witnesses, you don't believe the events.

A UFO flap originating in Gulf Breeze, Florida, near the Naval Air Station at Pensacola was just such an event, beginning at the end of Ronald Reagan's term in 1987 and continuing through the Bush 41 years into the early years of Bill Clinton's presidency. The Gulf Breeze sightings captured national attention and galvanized an entire community, which, at first, regarded the revelations of local building construction manager Ed Walters, who said he was the victim of attempted abductions by creatures in a hovering flying saucer, as the ravings of a crank until other witnesses came forward to substantiate the sightings that Walters claimed to be having. The Walters case gained national attention when the then head of the Mutual UFO Network, Walt Andrus, arranged for a photographic experiment using a version of a tamper-proof Polaroid camera that seemed to indicate that was what Walters was claiming to have seen was actually there.

The Night Of

The Gulf Breeze incident began quietly enough on November 11, 1987 when Ed Walters, working late at night at his desk at home looked out his window to see a strange light in the sky. It wasn't the moon, too close for that, and not a star, too bright for that. And it was the wrong color and in the wrong place in the sky. But the source of the light, whatever it was, was obscured by the branches of a tree on his property. Therefore, to get a better look, Walters walked outside, where, to his surprise, he noticed that the light seemed to get brighter. Worse, Walters would report to friends, he had the strange sensation that the light increased in intensity when he stepped outside to look at it, almost as if the light noticed him and reacted to his presence.

As Walters stood in his front yard watching the light grow in brightness, he also noticed that it seemed to be getting larger. That was when the realization struck him that it wasn't a motionless light at all. It was moving towards him. As the light descended, Walters could make out that it wasn't just a light, but an object with a definite shape: circular around the base with what seemed to be a turret on top, almost like a child's top.

Although transfixed at first, Walters realized that in order to corroborate his sighting, he'd need pictures of the thing, so he went back into his house to retrieve his Polaroid and then went back outside, where he snapped off a series of photos, pulling out the packets and watching the images develop. It was exactly as he saw it. Unmistakably a UFO with a type of checkerboard pattern along the base that looked eerily like windows. Next, he thought, if he could get better shots of the underside of the object, his photos would be even more compelling. So he walked farther out to the street and that's when, in an instant, he was hit by a beam of blue light.

At first, Walters felt the strangest sensation he had ever experienced. And then he realized that he was paralyzed in the spot where he was standing. He couldn't move out of the beam. The more he struggled, the tighter the lock seemed. There was an immediate feeling of panic, Walters later reported, but then he felt a calm come over him, almost like a reassuring message not to struggle. And that's also when he realized his feet were lifting off the ground. He had no feeling of flight or weightlessness, just the knowledge that his whole body was rising into the beam. And, almost like a sedative, a soft hum was playing through his ears. If he tried to move, it was no use. And then, he heard a dog's bark in the distance.

Walters could see through the blue light that a neighbor was walking his dog nearby. Did the neighbor see him, see the object over his head, see him rising from the ground? Suddenly, the beam snapped off and Walters dropped to the ground like a sack of potatoes and ran back into the house without even looking around. He still had the photos and camera in his hand.

The Next Morning

Would that it had been a nightmare, now washed away in the morning sun. But it wasn't. And Ed Walters had the stunning photos to prove it, detailed photos of a circular object with a turret on top and an articulated base, all of it lit from within. These were photos he had to share. But, even as he contemplated going to the local newspaper, the *Gulf Breeze Sentinel*, he had second thoughts. What if his construction management business suffered? After all, who would want to hire a construction manager who runs out into the middle of the street at night to snap off photos of a flying saucer? Then an idea struck him. Of course, these weren't his photos. They were taken by a friend, Mr. X. That's it, Mr. X. *He* took the photos and gave them to Walters to share because he wanted to keep his identity secret. And that's the story Walters told the *Gulf Breeze Sentinel*, confident that his own identity would be protected.

But he was wrong.

Other Witnesses

If Walters thought that the story of Mr. X and his dramatic photos would remain a local story, he would soon learn otherwise. First, other people in the area had also seen the object low in the sky, and they came forward. Then other newspapers in surrounding communities picked up the story and it began to circulate. Then it hit the wire services. Even people near the Naval Air Station reported having seen the object, which certainly would have run up the chain of naval command all the way to the Pentagon and the Defense Department, possibly to a UFO enthusiast like Ronald Reagan. It almost certainly would have been passed through intelligence sources to former DCI, Vice President Bush.

Under the weight of publicity, Ed Walters finally revealed that it was he who took the photos, saw the object, and had been trapped in its beam. Walters also revealed that his sightings had continued, had morphed into actual contact and that the contact was becoming increasingly menacing.

Contact

Not only was the familiar hum of the object that had initially trapped him in its beam become a recurring sound whenever the object reappeared, but it seemed as if those life forms on the object were trying to entice him to come aboard. They called him "Zehas." One night, actually in the predawn hours, Walters herd the incessant barking of a dog. Awakened, but still drowsy and suspecting nothing other than the dog might have been disturbed by an intruder, Walters staggered out into his backyard where he was met with a fearsome sight. This was no human intruder, but a child-sized creature with an enormous head only a few paces from his door and walking towards him. The creature seemed to be undulating in a strange way as if it was having trouble adjusting itself to Earth's gravity or atmosphere. Then it walked off in the direction of the woods. Walters attempted to follow it, but as soon as he walked out of the shelter of his doorway, he was trapped in the blue beam again, and paralyzed, until the creature disappeared into the woods and the beam snapped off, allowing him back into the house. This was unnerving, but it was not the last time he would see the creature.

A few days later, a now obsessed Ed Walters, looking for more proof of the existence of the object and of the creature he saw behind his house, saw the object hovering over a school a few streets away from his house. He grabbed his camera and walked to the school, where he saw the little creature again. But before he could snap a photo of the creature, the blue beam from the craft snapped on, caught the creature, and beamed it up.

Throughout the rest of 1987, Walters saw the craft on many occasions and finally began reaching out to experts to confirm what it was that was either tracking him or crossing his path.

The Experts

The experts in the field of UFO research had already heard the story of Walters's sighting even before Ed reached out to the research groups because after the wire services began circulating the story of Walters's encounter, it had become a national sensation. The Center for UFO Studies almost immediately debunked Walters's story because they deemed it too fantastic to be real. But Walt Andrus, the founder of the Mutual UFO Network, became very intrigued and sought a method to prove or disprove what Walters was reporting.

Andrus began by taking Walters's original Polaroid photos and turning them over to navy optical specialist Dr. Bruce Maccabee for analysis. Maccabee said that, given the nature of the Polaroid camera, it would have been nearly impossible for an inexperienced photographer, nay, photo hoaxer, to have staged these photographs. It is difficult to take double exposures using a Polaroid, but Andrus wanted to get further photo proof and provided Walters with special cameras, one of which was sealed with wax to prevent any tampering with the photo package, so that the veracity of Walters's story could be substantiated.

Debunkers Persist

The news that Ed Walters was taking photos of his encounters with flying saucers and little creatures was enough to arouse the interest of tabloid newspapers, who offered him money for his story. At the same time, the tabloids contracted with skeptics and debunkers to prove that Walters was hoaxing the entire thing. As the sensationalism about Walters's encounters spread, Walters' life itself became the object of scrutiny with negative stories about him turning up in newspapers around the country. To satisfy the doubters, Walters took multiple polygraph tests, which he passed, and underwent a psychological examination, which he also passed, as if one can pass a psych exam. He did not have any psychological problems, and he was not lying.

Pensacola

By now the sightings of unusual circular craft had spread to the naval base itself, prompting all sorts alarms from the base security. And because Walters was by now a national story, his house became a tourist destination for folks wanting their own UFO experiences and for people just wanting to see where he lived. Walters's private life was in turmoil, just as he feared it would be when he first decided to turn the photos over to the newspaper.

Debunking and Debunking the Debunkers

The intensity of interest in his house from tourists, who treated it as a Mecca for UFO visitations, finally drove Ed Walters away. He moved out and put the property up for sale, but it remained vacant for a significant amount of time before the new owners moved in. Then, according to stories told by the debunkers, the new owners were exploring parts of the house under the roof when they discovered a model of a UFO that looked just like the UFO Walters had photographed. "Hoax!" screamed the press. Walters was found out at last.

However, Dr. Bruce Maccabee became very suspicious of the convenience of the sudden discovery for the debunkers. He researched the entire "discovery" and here's what he found, telling his story on *UFO Hunters*. Shortly before the new owners moved in, a maintenance person showed up telling the owners that he was there to inspect the pipes in the attic. He left after purportedly doing his job. Then, after the new owners moved in, a reporter showed up, announcing that he wanted to interview the new owners and possibly take a walk-through of the property to see if there was anything of UFO interest for his story. On the walk-through, the self-described reporter led them right to the attic as if he knew where to go, where, lo and behold, there was a model of a UFO just like the one in the Walters photographs.

And when Maccabee reported the strange confluence of events surrounding the "discovery" of the model, the debunkers were completely debunked, and the Gulf Breeze sightings passed into history during the Bush administration.[7]

President George H. W. Bush on the 2016 Campaign Trail

If President Bush had been reluctant to talk openly about UFOs, a compelling statement he made at a campaign event for his son Governor Jeb Bush of Florida, who was running for the Republican presidential nomination in 2016, was probably more forthcoming than anything else he'd said. Asked by a reporter at a fundraiser for his son if any president would tell America the truth about UFOs, former president Bush said, and said it in the presence of his two sons and the assembled press corps, "Americans can't handle the truth about UFOs." If anything, this statement, while beyond compelling, also comports with other statements by folks who've claimed to have seen mentions of "UFO files" in government possession. According to one such former Army Air Force officer who later worked for Lockheed's top secret Skunk Works, the UFO files would never be released because if the truth were known, all the world's religions would collapse. This, more than anything else, points to the seriousness of America's history with UFOs and the national policy that derived therefrom.

Chapter 17

THE CLINTONS COME TO TOWN

Bill Clinton and UFOs

In 1999, the House of Representatives passed a bill that would have required across-the-board cuts to most federal agencies and departments, including the Department of Education, which President Clinton learned about as he was about to meet with a delegation of educators. The budget cuts also meant that teachers would probably not be getting any raises, something that irked Clinton to the point where, in an impromptu and unscripted remark, he said, channeling Ronald Reagan's famous statement, "If we were being attacked by space aliens, we wouldn't be playing these kinds of games." Funny how when Ronald Reagan said the same thing to the United Nations, folks commented that Reagan sure knew how to illustrate a point. When Clinton made his statement, Rush Limbaugh thundered into his microphone, "What's he going to do, arrange one?"[1]

Just three years earlier, while on a trip to Ireland where he was visiting a very troubled Belfast, Clinton read a letter he received from a child named Ryan, who had asked him about what he knew regarding stories of a UFO crash at Roswell, New Mexico. Clinton hadn't come to talk about UFOs. He was trying to make a point regarding how children can be victimized by political violence. In front of his Belfast audience, Clinton said to Ryan, "No, as far as I know, an alien spacecraft did not crash in Roswell, New Mexico, in 1947." But then he added, to the delight of his audience, "and Ryan, if the United States Air Force did recover alien bodies, they didn't tell me about it, either, and I want to know."[2]

Clinton did want to know, Webster Hubbell, Clinton's associate attorney general, wrote in his own memoir.[3] As AAG, Hubbell claimed that

President Clinton asked him to find out all that he could about two things: who killed JFK and what the government knew about UFOs. He reported to the president after being stonewalled by the relevant agencies that there was a secret government that closely holds secrets to which the president doesn't even have access.

However, in 1993, the year Clinton was inaugurated, and as pressure was mounting on the CIA to confirm that it had followed UFO stories, Director of Central Intelligence James Woolsey "ordered another review of all Agency files on UFOs. Using CIA records compiled from that review, the study he ordered traced CIA interest and involvement in the UFO controversy from the late 1940s to 1990."[4] The report reveals that despite official statements to the contrary regarding the government's interest in what happened at Roswell, the air force did, in fact, open up an investigative unit, known first as Project SAUCER and then Project SIGN, to "collect, collate, evaluate, and distribute within the government all information relating to such sightings, on the premise that UFOs might be real and of national security concern."[5] The CIA report reveals that the agency followed the air force investigations into UFOs, especially the invasion of Washington, DC airspace in 1952 and the reports of pilots and radar operators who said they were seeing images they could not identify and did not seem to represent conventional aircraft. The report continues to explain the nature of CIA involvement in UFO cases over the decades and, without making the point explicitly, implies that despite any official denials, and, moreover, very relevant to the background of George H. W. Bush, the agency was very active in UFO research, had compiled voluminous files on UFOs, and had much to share with any American president, should the DCI have reason to do so. The report must have been a real eye-opener to President Clinton, not so much because of what truths behind cases it revealed or did not reveal, but simply that the files that the president had dispatched Webster Hubbell to locate, files the Hubbell said had been kept from him, now, albeit sanitized, were available to him.

According to Jon Austin in the UK's *Express*,[6] as far back as Bill Clinton's first term in office in 1993, Hillary Clinton was talking up UFOs with Laurence Rockefeller, who funded initiatives to research UFOs almost as a way to counter what the Condon Report had argued as well as making its own case for the study of the possibility of extraterrestrial visitations.

"From early 1993 the businessman began a lengthy approach to Bill Clinton for disclosure, including files held by the CIA, in what became the Rockefeller Initiative."[7] Further, Austin writes, "During the Clinton family seven-day vacation in August 1995 at the Rockefeller Teton Ranch outside of Jackson Hole, Wyoming, the 85-year-old billionaire privately briefed the President and First Lady on UFOs and his hopes.

"No one else knows exactly what was said, but at the time Mr. Clinton, who had a personal interest in UFOs and was frustrated at the lack of information he could glean on it, was carrying out a review of how the Government handled confidential material."[8] For a scientific review of the Laurence Rockefeller conference on UFOs at Pocantico Hills, the basis for the Rockefeller conclusions about the need to research the UFO phenomenon further, see Peter A. Sturrock's analysis *The UFO Enigma*, in which Rockefeller himself explains the purpose of his initiative.[9]

The Clinton perspective on UFOs, however, took a dramatic turn in March 1997, the time of the Phoenix Lights, the arrival of the Hale-Bopp comet, and the Heaven's Gate mass suicide, when President Clinton himself would become involved in the swirling cover-up of the most media covered UFO appearance since the 1952 invasion over Washington, DC.[10]

The Phoenix Lights

On March 13, 1997, the stage was set for thousands of witnesses across the south west scanning the sky for the appearance of the Hale-Bopp comet—for comet watchers, a once-in-a-lifetime event. But it wasn't just a comet that residents of Phoenix saw on that warm March night. The story actually began before 8:30 p.m. when witnesses in Henderson, Nevada, at the Arizona border reported a strange sight, six orange lights in a triangular pattern slowly moving toward the southeast and making no sound except for what the observer said was the wind. About fifteen minutes later another witness, this time behind the wheel of his car, saw a formation of lights heading his way as he drove north. He headed home, retrieved binoculars, and kept watching the lights as they headed south and west towards Phoenix. And still another person, this time actor Kurt Russell at the controls of his private airplane, who was flying his son into Phoenix, also saw the strange lights in the sky, reported his sighting to air traffic control, and was told they could not identify the source of the lights ahead of him and could not see "anything out of the ordinary" in his area. In May 2017, Russell, appearing on the BBC's *One Show*, while on a promotional tour for his film *Guardians of the Galaxy Vol. 2*, told host Gyles Brandreth that when the stories of the mysterious pilot who reported the lights began to circulate in the news, he realized they were talking about him. But he said that he had forgotten about sighting the lights at first and only realized what the incident was about years later. He told his interviewer, "The fascinating part to me is that it just went, literally, out of my head." Russell's revelation that he was the mysterious pilot who'd spotted the lights broke a twenty-year mystery concerning the identity of the witness.[11] Interestingly, Russell's statement that the lights didn't seem to intrude that heavily into his consciousness was similar to statements made by other close-up witnesses that after observing the lights, they simply went back about their respective businesses. Russell's description, and the descriptions of the triangles to which the orbs seemed to be attached, were similar to the giant triangles witnesses saw over the Hudson Valley and Indian Point almost fifteen years earlier.

Within minutes after Russell's sighting, folks just north of Phoenix began seeing the lights off in the distance to the north-northwest, heading their way. It passed right over private streets, possibly as low as 100-150 feet, and traveling slowly toward the south. Then it seemed to stop dead in the air. It hovered. Then it picked up motion and continued through the narrow valleys toward the outskirts of the city suburbs where more people, some on the roads, some standing on their balconies, spotted the object, which many people described as unearthly.

One of the witnesses was Arizona governor Fife Symington, who said in 2007 and on *UFO Hunters* that despite what he said at his faux news conference when he had his chief of staff dress up in an alien garb and show up as the explanation for the lights, Symington himself saw the 8:30 lights from his own backyard and said they were not a conventional craft because the object he saw flew too slowly for an airplane, did not have the thwapping sound of rotor blades like a helicopter, and was completely noiseless.[12] A plane would have stalled at the speed it was moving, a helicopter would have shown distinct navigation lights and made a noise, especially at the low altitude at which it was flying, and the lights made the craft larger than a football field. Also, Symington said that as commander in chief of the

Arizona National Guard and Air National Guard, he called Luke Air Force Base and was told no military craft were in the sky at 8:30 over Phoenix. Thus the lights were not confirmed as military by the Air National Guard.

About two hours after the first sighting of the lights, a second set of lights flew over the same area, causing even more confusion and frenzy than the first set. Were the 10:30 lights over Phoenix a different object, the same object returning, or flares from air force or National Guard planes scrambled to establish a conventional explanation for the 8:30 lights? At first, when the 10:30 lights appeared, residents called Sky Harbor air traffic control again and were told that no planes were picked up on radar. Residents also called Luke Air Force Base again and were told none of their planes were in the air. However, after the initial responses from Luke AFB, the air force said that the lights were ground illumination flares from a flight of A-10 Warthogs over the Goldwater test range. Months later, the Maryland Air National Guard announced that there were planes in the air on the night of March 13, 1997, from the 104th fighter squadron flying out of Davis-Monthan Air Force Base in Arizona as the unit that dropped LUU-2B/B ground illumination flares in an exercise over the Goldwater test range. CIA UFO consultant Dr. Bruce Maccabee conducted a triangulation analysis of the events and said that it was indeed likely that the second formation of lights were flares that could be seen even though they were many miles south of the Phoenix area at 10:30. However, photo analyst Jim Dilettoso of Arizona's Village Labs said that after he had performed a spectral analysis of the lights, he determined that the colors of the lights were inconsistent with flares.

Witness and award-winning filmmaker Dr. Lynne Kitei has said that the 10:30 lights were not flares and that subsequent flare-dropping demonstrations by the air force to prove that flares were what people in Phoenix saw on the night of March 13, 1997, showed, again through an analysis of the aerial performance, the colors and the residue actually proved that what witnesses saw were not flares. The controversy over the nature of the lights still remains today.[13]

This was the night of the comet, an event that thousands of sky watchers were set to enjoy even as members of the Heaven's Gate cult in San Diego, led by Marshall Applewhite, prepared to take a lethal overdose of sedatives to commit a mass suicide. But what sky watchers saw that night along a corridor running from Nevada to the Mexican border across Arizona was something far more than they bargained for. They saw a triangular or arrow-head formation of orange/red circular orbs, self-illuminating lights moving slowly and silently overhead, low to the ground, which seemed, at times, to be hovering, not dropping, even though there air currents aloft. Witnesses to the 8:30 lights say they actually saw a rigid object holding the lights in place, describing it as a V-shaped craft or a boomerang almost a mile wide and stretching across the entire sky. The people who watched the spectacle were actually in shock, commenting on the local and even national news that it was nothing like anyone had seen before.

The *Arizona Republic* newspaper's Richard Ruelas, who has been covering the Phoenix Lights story from its beginning, said, "A lot of times UFOs are seen in isolated and rural areas, usually by a single witness or a small group. However, this was the first mass sighting in a heavily populated area."[14] Even more importantly, the witnesses were not just folks out for a stroll at night. These were, in many cases, professional individuals, doctors and former military, who lived in the vicinity of Sky Harbor Airport and knew what conventional aircraft, including helicopters, looked like in the night sky and how they performed.

"One of the big confusions," Ruelas explained, "was that there were two separate events. At around 8 to 8:30 folks see a V formation. At around 10:30 there are hovering orbs over the western part of Phoenix." He continued, "Most people on the ground who saw the first lights saw the flying V."[15] This seemed to them to be a real anomaly.

The mystery of the flying V prompted City Councilwoman Francis Emma Barwood on May 6, 1997, to ask what these lights were and why they caused so much confusion, a question she raised with the mayor and city council when she asked for an air force explanation for the phenomenon and to express her query as a public safety concern because of reports of an unidentified object over a populated area. She had asked, "What's hovering over the city?"[16] Ruelas said that when Barwood raised her concerns, it meant that a public official requesting an expenditure of public resources because of a possible UFO had the immediate result that, "it gets her laughed at." *Arizona Republic* reported that Barwood didn't want to be known as the "UFO candidate," but that description follows her everywhere.[17]

Barwood told James Fox on *UFO Hunters* that she never said the word "UFO," it was not mentioned at all.[18] All she asked about was whether any official investigation revealed anything about the nature of whatever was in the sky over Phoenix because it was close enough to the airport, as well as over a densely populated residential area, that in her mind it touched on the area of public safety. She never said, "UFO, flying saucer, extraterrestrial craft." She never said anything like that. Barwood said, "I couldn't understand why no one wanted to find the truth. Why didn't anyone want to know what it was that night?"[19] But she was ridiculed in the *Arizona Republic*, which, she said, "ran a cartoon of me with flying saucers over my head and another with a switch embedded in my head." Newspapers and even some of her colleagues on the city council became very derisive towards her, some referring to her as "beam-me-up Barwood" even though she never suggested any extraterrestrial explanation for the lights over Phoenix. She continued that she started getting phone calls at seven in the morning, calls which didn't stop until after eleven at night. "Over the summer, I tried to call back everybody, over 700 people," she said. Asked if she pursued any independent investigation on her own, Barwood said she referred all the phone calls she received to Governor Symington.

Barwood also talked about Governor Symington's news conference early on June 19, 1997, when, pressured to find an answer to the sightings, he promised to "look into the issue." "We're going to find out if it was a UFO," he announced. That afternoon, the governor called for another unscheduled press conference, at which he said he had found the culprit behind the whole incident. "I will now ask Officer Stein to escort the accused into the room so we may all look upon the guilty party," the governor assured the assembled media representatives. Barwood said that at first there was excitement in the room because he said, "he was going to look into this." But when the governor's chief of staff walked into the room dressed in an ET costume, large head, long fingers, and a long shining nightgown, "we realized," Barwood said, "that this was a big joke. Given the thousands of witnesses who wanted answers, the governor was laughing at the entire incident." But, Barwood said, "this was not at all like Fife. He doesn't have a sense of humor." As the costumed alien began to take off his false head, Symington told reporters that "this goes to show that you are all you are

entirely too serious." The laughter resounded through the room. But there
was no laughter from the witnesses.

"Was it your perception," Barwood was asked, given all the media and
derision surrounding the lights incident, that "there was something organized against you because you brought up your questions at a city council
meeting?"[20]

"It seems that way," she said. "Many people wanted to know answers,
but never did I have any negative comments from my constituents or
anybody in the city or anybody in the state. The negativity came from government officials."[21]

Witness Terry Mansfield said that she saw a boomerang-shaped craft,
an actual craft she could have almost touched because it was so close to her
house. She was hosting a meeting in her house when one of the volunteers
in a project she was working on pointed out the window and said, "Oh, my
God, look what's outside."[22] Terry said the group all looked straight from
her porch and realized that the sky seemed to have gone black because there
were no stars. They had been blacked out by something. They realized also
that they were looking at a black, shimmering, satin-like material that comprised a rigid shape. "The craft was so big," Terry said, "that I couldn't see
the back or front." The craft, she said, spanned the entire horizon, completely blocking out the sky overhead. She described the underside of the
craft as undulating and fluid, that was so close as it floated overhead without a sound as something she could reach out and touch. Terry, who had
been in an air force family for over twenty-five years, said that she had seen
every type of military aircraft in service. The craft resembled nothing she
had ever seen before. If anything, it was the size and the undulating nature
of the craft's skin, as well as its noiselessness, that impressed her most, as it
did her friends who witnessed it with her. Yet there was another aspect to
this craft's appearance effect on her group. After being struck by the proximity and awesomeness of the spectacle, the women went back inside to
their meeting without saying a word to each other about their shared experience. It was if the women had come to a collective agreement not to talk
about what they had seen. Maybe it was the nature of the spectacle, Terry
wondered, but it also seemed as if they had silently agreed that they were
not supposed to talk about it. For his part, Pat Uskert of *UFO Hunters* was

struck by the number of identical witness descriptions, each corroborating the other about the characteristics of what they saw.[23]

For the record, Governor Symington said, "I saw a large velvety illuminated triangular shaped object."[24] But why did the governor mock the entire event the first time he promised to investigate it? "First of all," Symington admitted, "the world was descending on us. Media from everywhere appeared in Phoenix wanting to know what's happening. I made the decision at that point to spoof and to lighten the atmosphere. But I didn't realize that I was offending so many people. I was simply trying to lend a little levity to an atmosphere that I thought was getting too hyped up, too over the top, too much hysteria."[25]

However, in light of his promise initially to pursue all avenues of investigation into determining what caused the Phoenix Lights, what avenues did he take and what did he learn? He said that he had ordered his staff to contact the air force at Luke Air Force Base and to call the Department of Public Safety and the general in charge of the Arizona guard to find out what it was. "But the air force totally blanked us. They just said, 'we have no comment.' That was it."

Wasn't that a non-answer, a refusal to confirm or deny anything, an admission that it's something you can talk about?

Symington again revealed that for politicians, a "no comment" means that there is actually something you can't comment about. It's a way to suggest that something you can't talk about does exist. The air force knew there was something there, but would not talk about it. Thus, for the first time on television, a high government official admitted that his inquiry into the Phoenix Lights indicated that there was something the military was keeping secret even from the civilian official in command of the military in Arizona.

Symington has also said on camera that after the whole incident was wrapped up, he received a pardon from President Clinton, even though the pardon was for bank fraud, the timing, after Symington's UFO news conference made the true reasons for a pardon very suspicious. Bill Clinton, according to all reports about the night of the Phoenix Lights, actually disappeared from view. He had been staying at golf champion Greg Norman's house when reporters were told Clinton suffered a bad injury to his knee and had to be laid up. Did he really or was he huddling in emergency session with aides and military officers about the breaching of restricted airspace over Phoenix? We also know that the air force scrambled a flight of F-15s that night, one of which captured video of the 8:30 lights, which video was removed from the plane and taken to Washington. Thus, one has to ask, what did President Clinton find out about the Phoenix Lights, and why did he pardon Governor Symington in 2001, years after the faked news conference? Was it because, according to a suggestion from MSNBC's Rachel Maddow, Symington once saved Clinton from drowning when Clinton was a college student or was it that Symington deescalated the growing controversy over the Phoenix Lights? Moreover, what did Symington, or President Clinton, for that matter, tell John Podesta and First Lady Hillary Clinton, the Democratic Party's 2016 presidential nominee who has promised to reveal what the government knows about UFOs as long as it doesn't compromise national security?

Fife Symington was not just the governor of Arizona. He hailed from an important political family in which his cousin, Stuart Symington, former secretary of the air force during the immediate period after the crash at Roswell, was also the US Senator from Missouri and candidate for the Democratic presidential nomination against JFK in 1960. Stuart Symington would have been the specific individual in office as the Roswell cover-up was implemented, would have been administratively involved in the beginnings of Projects Sign and Blue Book, and would probably have been in the loop in 1952 as the boss of air force generals Sanford and Ramey during the UFO invasion of Washington, DC, airspace.

Given that background, we can only wonder what secrets his cousin Fife Symington, also an air force officer, might have been privy to. That said, Symington admitted on camera that he had seen the giant triangle that was the source of the Phoenix Lights at 8:30 that evening, telling people that he saw a large triangle fly so low over his backyard that he could have hit it with a rock. He noted that through the orb-like lights on the tips of the triangle, he could see wavy starlight as if he were looking at it through translucent paper. It was, he said, a stunning sight and something he believed that did not originate from this planet.

Was this the incident that brought Bill and Hillary Clinton into the loop? And what about Hillary's flirtation with the ghost of Eleanor Roosevelt and with the secret of UFOs that would still linger over the Clinton legacy as she mounted her own campaigns for the presidency in 2008 and especially in 2016, when opponents from as far away as Montenegro, Moscow, the New York office of the FBI and its retirees, a former New York City mayor and her quondam unofficial opponent in her first senatorial election, and the candidate himself from inside the high castle of Trump Tower resurrected the Salem Witch Trials against her as his acolytes chanted from the hymnal over the plaintive braying wail of a country music singer, "Lock her up?"

But wait, the plot sickens.

Chapter 18

PRESIDENT GEORGE W. BUSH, VICE PRESIDENT CHENEY, AND THE UFOS

Vice President Cheney and the UFO Files

As the story goes, when a jovial George W. Bush was asked whether the government would ever come clean about what it knows about UFOs, Bush jerked his thumb at Cheney and told the audience that if there were any bits of information about UFOs floating around inside government files, Cheney would be the man to talk about it. Why was that? For his part, Dick Cheney told UFO historian Grant Cameron in a 2001 radio interview that "if I had been briefed on it, I'm sure it was probably classified and I couldn't talk about it."[1] Thus, he could not even affirm that he had been briefed on UFOs.

However, presidential candidate George W. Bush said in 2000, in answer to a question posed to him in Arkansas, that it would be Dick Cheney's "first White House job to get the answer to the UFO mystery."[2] And after the Bush statement about his running mate, Cheney showed up in Roswell, New Mexico, a few months later, almost as if teasing the press about UFO disclosure. Further deepening the mystery, UFO presidential historian Grant Cameron points out, "Almost as if it had been written by a Hollywood screen writer, only six days before Cheney arrived in Roswell, 30 cattle were found dead under mysterious circumstances on a ranch outside of town. Papers as far away as Boston, Massachusetts covered the story of the 'mysterious mass death.'"[3]

Stories of cattle mutilations—the removal of soft tissue such as material from the gums, genitals, and tissue from inside the eyes—were as much a part of UFO lore as abductions. Cattle ranchers and farmers have been reporting since the late 1960s about the discovery of dead cows that showed no other symptoms of disease, but which had been drained of blood and had certain organs meticulously excised from the cow, whose carcass suffered no desiccation, no source of food for scavenger birds or insects or from other animals such as coyotes. If cattle in the American southwest had died from radiation poisoning, which was an ongoing theory for decades, veterinary necrologists might have found residual radiation in the animals' tissue samples. But no radiation results were returned. Thus, the mutilations were a mystery, especially occurring just days before Cheney's arrival in Roswell. A veterinary specialist did suggest, after Cheney had left Roswell, that the cause of the cattle deaths was poisoning from a chemical found in some of the hay the cattle were eating.[4]

Letters to Vice President Cheney during the early months of Bush's first term, urging him to release whatever government files on UFOs that he could, went either unanswered or, for those that actually reached his office, were given polite responses that did not use the term UFO. Simply stated, if there were a hope that someone in the administration, either the vice president or Secretary of State Colin Powell, would respond to the UFO inquiries, that hope was dashed.

In 2002, however, there was a renewed interest in the forty-seven-year-old crash at Kecksburg, Pennsylvania, when former Clinton White House Chief of Staff John Podesta, then having left the White House, came out in favor of the disclosure of all government files on UFOs by supporting Leslie Kean's Coalition for Freedom of Information lawsuit regarding release of NASA's Kecksburg files. When he was at the White House, according to Leslie Kean at the *Huffington Post*, he had advocated for "Executive Order 12958 which declassified broad categories of information formerly withheld on the grounds of national security."[5]

When the Coalition for Freedom of Information lawsuit, now joined by a former White House Chief of Staff, reached the US District Court in Washington, DC, it would have been unsurprising if an intelligence-hawk like the vice president did not get the news of the pending court ruling. In that ruling, in Kean's words, "the exasperated Presiding Judge, Emmet G. Sullivan, declared that 'heads should roll' at NASA and that NASA's case was a 'ball of yarn.'" He approved a settlement that required NASA to comb through hundreds of documents in specified locations, and provide them to journalist and plaintiff Leslie Kean, under the watchful eye of the court.[6] But according to former NASA employee John Scheussler, there was a group at NASA, as there are in other federal agencies such as the military, that had access to what he referred to as Special Access Projects in a Special Access Program that was very limited. Scheussler remembers that certain types of information, particularly concerning UFOs and similar anomalies, were the province of the Special Access Program and were not shared even at the highest levels of NASA.[7] Thus, it would not be far-fetched to find out that there was no information at NASA, and what might have been there had already been scrubbed by those in the Special Access Projects Group.

Leslie Kean conceded this possibility when she wrote about the documents turned over to them by NASA. "The resulting documents did not include one iota of information relating to the Kecksburg case, despite an earnest and thorough effort by NASA staff. Three hundred boxes were searched, on top of the hundreds of pages released during the court proceedings prior to the settlement. Yet they revealed nothing about the case at hand.

"Could these be the 'UFO files' that Podesta was alluding to in his tweet? Might there be others as well, stored in locations inaccessible to the FOIA in violation of the law?"[8] And, even more intriguing, would any of the folks in charge of Special Access information actively intervene in purging files once the court's decision had been handed down? The lawsuit for the files had been working its way through the judicial system before it reached the hearing stage at the 2nd District. NASA attorneys, perhaps aware of the existence of special access personnel, would have most certainly briefed their superiors at the agency regarding the progress of the suit and, more than likely, their predictions regarding the ultimate court's ruling. The files could well have been scrubbed in advance of the judge's decision.

In all fairness to the Bush administration, this was a presidency consumed by war from its first nine months in office right up until the election of President Barack Obama. Given the strategy of the think tank groups to which Vice President Cheney belonged, the national security advice both Cheney and Bush were getting from the intelligence community, and the war communiqués the administration would be receiving in the immediate days after the invasion of Afghanistan, it's not surprising that the subject of UFOs, even though reported sightings were taking place not only in the

United States, but around the world, would be dismissed as not particularly relevant in light of the administration's all-consuming focus on war. Whether one is conservative or progressive, we can all appreciate President Obama's admonition that he would not talk about UFOs, especially in light of the commander-in-chief's hardest and most painful decision: sending young people in the military into harm's way during an otherwise useless conflict where they could be injured for life or killed. Thus, during Bush's wartime presidency, even the news that Podesta and Leslie Kean had prevailed in their lawsuit against NASA barely made the newspapers in the United States even though tabloids in other nations took notice of the court's decision and NASA's responses thereto.

John Podesta and the UFO Files

John Podesta's interest in the UFO question and how his inquiries and legal activities glided stealthily beneath the frenzy of the Bush administration is best explained by his immediate history with the Clintons and their fascination with the UFO question. He certainly would have known about Hillary's predilection for investigating around the edges of the UFO issue, about her conversations with Laurence Rockefeller, and about the UFO-themed books she was alleged to have read, including those titles about what the government is reputed to know about the origin and intentions of UFOs. Perhaps Podesta was coordinating the communications traffic on the night of the Phoenix Lights when President Clinton just about disappeared from view when he was visiting Greg Norman's house and claimed he'd hurt his knee. Perhaps Podesta was in the loop when the media began besieging the White House with questions about what the government was doing when the 8:30 and 10:30 lights floated over Phoenix and the city's Paradise Valley, causing a commotion throughout the American southwest.

Other question arise as well. Was the White House even peripherally involved in encouraging Governor Fife Symington to hold a fake news conference to "lighten the mood" regarding the invasion of Arizona's restricted airspace or was it the secret government that had read the White House in on its plans? Did anyone on Clinton's National Security Council see the video taken by one of the two F-15s launched from Luke Air Force Base to surveil the triangular object to which the lights were seemingly affixed? Did Governor Symington relay a report of his firsthand eyewitness observation of the floating giant triangle to anyone in the Department of Defense and thence to the White House? And was there any discussion involving Podesta even tangentially regarding the presidential pardon for Fife Symington?

Simply stated, we don't actually know what went on in the White House, much less what might have gone on that involved John Podesta in an administrative or even a personal role regarding what surely must have been, at the least, heightened interest, and, at the worst, a military alert when the lights showed up over the American southwest. But if Podesta was in the loop of any UFO discussions, because of any national security clearances he might have had in order to discuss any classified issues with the president, then it's likely that as White House Chief of Staff and one of the chief counselors to President Clinton, Podesta would have been a party to even the most benign of discussions about the Phoenix Lights, Clinton's whereabouts that night, the mass suicide in San Diego, and the international media interest in the events over Phoenix.

The amount of interest that the entire Phoenix event generated, interest that has continued for the ensuing twenty years and was only recently reinvigorated with the Kurt Russell as the mysterious pilot revelation, then it stands to reason that Podesta would have been staring a swirling mass of concern buried within a black box of disinformation, misinformation, or simple denials in the face of overwhelming questions. That experience would not have dissipated upon Podesta's leaving the White House at the end of Clinton's term, especially when, only eight months into the new administration, the World Trade Center and the Pentagon were attacked by terrorists using commercial airliners as missiles. Podesta, although he wasn't a UFO advocate per se, nevertheless said he believed in the necessity of governments' being as transparent and open as possible to their constituents. Thus, given the secrets that he knew from the Clinton administration, and given his own restrictions and constraints on revealing information while in the Clinton White House, it must have been liberating, to say the least, that he could approach the issues of government transparency and open records once he left the White House after the Bush 2000 election.

All of this speaks to the rationale for Podesta's joining Leslie Kean's Coalition lawsuit to release the Kecksburg files that she and Podesta believed were being kept secret by NASA, whose personnel were part of the retrieval team of the Kecksburg object that landed in 1965. And when the time came for him to come on board Hillary Clinton's 2016 campaign for the Democratic nomination and then the presidency, it was almost a foregone conclusion that the interest in the UFO community would be focused upon the Podesta email cache hacked by the Russians and released by Wikileaks. Those leaked emails and Hillary's teasing about UFOs would come to form one of the conspiracy threads of the 2016 election.

Chapter 19

THE OBAMAS AND UFOS

Perhaps the most telling remarks President Obama made to late-night television talk show host Jimmy Kimmel in response to his question about UFOs was, "they asked me not to talk about it." Who is the "they" and what is the "it"? But, despite his denials, at least three major UFO stories welcomed Barack Obama as he completed his oath to protect and defend the Constitution and entered the Oval Office.

The Stephenville Lights

We know that when one president leaves office at the end of a term, the president leaves a message for the successor in office, a message, reportedly, detailing the issues likely to confront the new president and, perhaps, even some suggestion for paths to take regarding those issues. So may it have been as George W. Bush was preparing to leave office in 2008. While he was facing concerns over a bursting bubble in the American housing and mortgage markets, as well the entire underpinning of the financial system that supported those markets, there was another breaking story in 2008 that almost surely captured President Bush's attention and that greeted the new President Barack Obama when he took office in January 2009, and that was the entire incident called the Stephenville Lights, UFO sightings and interactions with F-16 interceptors that took place very near President Bush's ranch in Crawford, Texas.

Although the earliest stories of strange lights in the skies over Stephenville began at the end of 2007, in 2008, the story of a possible UFO flap near Bush's ranch was picked up by the Associated Press. The intriguing story told of multiple witnesses who viewed a strange light over the ranch lands in Erath County centered around the small city of Stephenville. What made the story more credible than a single-witness sighting was the report by the local constable, a county law enforcement official, who said that not only did he see a large, brightly lit wingless object while he was behind the wheel of his police unit, he had photographed it with his dash camera, something that a local police sergeant, who was riding shotgun with him that night, also confirmed.[1] And a local pilot also said that he saw a large, well-lit, circular object in the sky. All in all, over two hundred people came forward to report seeing constellations of lights in the sky that they couldn't explain, which were followed by a formation of F-16s.

The story began to spin nationally when, after the Associated Press picked it up from the *Empire Tribune*, the local Stephenville newspaper, not only did other major media send some human interest—read "news of the weird"—beat reporters to the town, other UFO researchers flocked to Stephenville to interview witnesses, at which point a fight broke out between different UFO researcher factions, each seizing on its own witnesses to tell its version of the story. And as with many stories of this type, there were those looking to cash in on the media hubbub and see what they could scrape up from the free-spending reality and news shows that were turning up looking for witnesses, anyone with a story, along Main Street.

Although the hype threatened to become overwhelming, there were some very credible stories, and even an eerie video, reported by some witnesses. In fact, the video, assuming that it was real and not faked, was so troubling that if there were still an investigating arm of the government for UFOs—and believe us, there still is one—then the video and another witness description of an actual craft hovering over a local railroad crossing would have been enough to send a classified memo right up the line to the office in charge of Special Access Programs. Like the Phoenix Lights ten years earlier, there is a shock value to high-profile UFO sightings. When stunning witness reports surface in the news, a chillingly credible video from Thanksgiving 2007 turns up online, and a series of formulaic denials from local military officials at nearby Carswell Air Force Base about unexplained lights overhead followed by reports of F-16s scrambling to chase them turn up in the news, especially the international media, the president gets the memo whether POTUS wants it or not. In the Stephenville incident, only miles away from the president's ranch in Crawford, Texas, you can bet that an internal Secret Service alert circulated in the West Wing.

The eerie Thanksgiving 2007 video was taken by local resident Margie
Galvez, who wanted to find out what animal, assuming that it was an animal, was coming onto her property at night to make off with her chickens.
Was it a poacher? Was it a predator who left no trail behind? There were no
footprints, no signs of a struggle, nothing. What was it? To catch the culprit, Margie set up a small night-vision surveillance camera looking into
her backyard to see if anything turned up. But what she found in the murky
images not only surprised her, it added a very chilling aspect to the UFO
stories circulating around town. The grayscale video, picking up the heat of
objects against the background heat of the air as opposed to the light from
the objects, showed her chickens scratching and pecking at the ground. At
first, when Margie viewed the footage, she saw the lights of what looked
like animal eyes off in the distance, then a mysterious white light—which
meant that it was hotter than the surrounding air—suddenly shot down
from the top of the frame and seemed, just like a searchlight, to be scanning
the ground for something. Then it simply snapped off.[2] But what bothered
Margie the most, besides the obvious concern about what was shooting the
beam down from above the frame, was the nature of the beam itself. Why
was the beam dissipating before it hit the ground? It was an expanding cone
of light swinging back and forth across the frame like a pendulum, but
never touching the ground. It was clearly looking for something. Margie
Galvez never saw the object that projected the beam, but another witness,
this time in broad daylight, claimed she did.

This witness, who remains nameless, came forward in spring 2008 to
describe an event she observed when she was driving her teenage daughter
and her school friends home from cheerleading practice one afternoon outside of Stephenville. As the witness and her pickup truck full of teenagers
approached a railroad crossing, they were startled by a beam of light coming down from the sky and apparently scanning the ground for something.
When they looked up to find the source of the light, they were stunned
to see a large, larger than a conventional airplane, rectangular object with
no apparent engines or wings seemingly hovering above the ground. It was
flat, as if a floating platform, making no sound whatsoever, and it had portholes in the side and lights at either end. Pilot Steve Allen might have seen
the same thing, a large rectangular object with red, yellow, and green lights
at either end and traveling across the sky towards Stephenville at about four

thousand miles an hour. It was not a plane and Allen could not account for what it was.

Eyewitnesses notwithstanding, the entire incident became even more convoluted when other witnesses said they saw about ten to fifteen F-16's roaring across the sky that very same night. But calls to the local air base at Carswell got blanket denials of any aerial activity that night, that is, until the story hit the national and international wires about the flights of jets following a large, noiseless, well-lit object. Then, suddenly, the public information office at Carswell said, "wait, there was a training exercise that night." But here's the problem: Carswell was the local base and some local residents of the Stephenville area worked at the base. Other residents were friends of folks who worked at the base. All of them reported that had there been a big training mission involving ten to fifteen F-16s, the activity at the base would have been noted all over town. There would have been maintenance conducted on the planes, fueling would have taken time and would have been reported, and officers who worked at the base would have been called in. None of that happened. Therefore, the story of a training mission on the night the jets were in the air seemed to ring false to the folks in Stephenville.

After the stories of ongoing UFO sightings no longer made the local headlines, the real background grind work of analysis began. First, there were the flight logs from the various military areas around Stephenville, which showed normal air traffic into the civilian airports and no abnormal military traffic. However, examinations of various radar tracking records from the nights in question in early 2008 through the spring of that year did show objects in restricted air spaces that could not readily be identified. If commercial, there were no transponder readings. If military, flight logs did not support any claims of military flights that night. But because there were anomalous objects picked up on radar, clearly there was something in the skies over Stephenville, reports of which most assuredly went to President Bush because of the proximity to is Crawford ranch and thence to the new president, Barack Obama.

The Needles, California, Crash and the Men in Black

He called himself "Riverboat Bob," a local resident who plied the waters of the Colorado River outside of Needles, a river whose bends and banks he knew very well. At 3 a.m. on May 14, 2008, he saw a flaming turquoise object streaking out of the sky and hitting the ground along the banks of the river. As Bob pulled his boat up towards the bank to get a closer look, thinking it was possibly a meteor, albeit one with a very strange color, the sky above within minutes was filled with double-rotor helicopters—which looked military to him—and a large helicopter that had a sky crane attached. The sky crane hoisted the object, which was about thirty feet long and bulky and looked like a tanker truck, and, with the object still burning with a bright turquoise glow, flew it away. But Riverboat Bob wasn't the only witness to the object's falling. This was a multiple-witness sighting, which lent more credibility than if the only witness was a guy who lived on the river. Therefore, during that early morning darkness in the final year of the Bush administration and as both Senators Barack Obama and Hillary Clinton were squaring off in the Democratic Party primary election, another UFO flap broke across the western US like a clap of thunder.

At the same time as Riverboat Bob saw the object, a retired LAPD captain, airport security chief, former operations head at Los Angeles International Airport, and current Needles resident Frank Costigan walked out into his back yard to find his cat. He looked up at the nighttime desert sky and noticed a turquoise object fly overhead and pass into the distance, where he saw it descend toward the river. He thought he would soon hear the crash as the object fell, but he heard nothing.[3] The very next day, residents in Needles saw black unmarked utility vehicles and a large truck with a dome on top cruising through town, apparently looking for something in a way that seemed menacing to the townsfolk. The vehicles looked like a government convoy, but where was it from, what was it doing in Needles, what was under the dome, where were they going, and did the appearance of the vehicles have any relationship to the object that Costigan and Riverboat Bob saw? And then Coast to Coast AM radio host and KLAS-TV news correspondent George Knapp came to town to find out more about what had happened.

Knapp reported that he was confronted by the—for want of a better term—men in black, whom he confronted in return, asking who they were and where they were from. Eventually he found out they were from Area 51, the place of one of Knapp's most important stories about Bob Lazar and Lazar's stories of alien propulsion devices that could stop time and reverse engineered flying saucers hovering over the Nevada desert at night.

Why would the residents of Needles be intimidated by a military unit from Area 51 just a day after an otherwise innocuous incident along a secluded riverbank? And why would a local radio station manager who was reporting the story be, himself, threatened by this group of armed US government operatives? The events surrounding this incident in the desert town of Needles were very much like the events in Kecksburg, Pennsylvania, over forty years earlier: a mysteriously glowing cylindrical object making a controlled crash landing, military units retrieving the object, and military personnel intimidating into silence anyone claiming to be a witness. Who ordered this and who is still keeping this secret?

Yet despite the secrecy, the number of witnesses who saw the large helicopter lifting something off the ground and escorted by a number of other twin rotor helicopters indicated that this was more than a simple exercise. The military was removing something its personnel had expected to find there because of the speed at which they arrived on scene.

David Hayes, the owner/manager and on-air host of Needles KTOX radio station, in the wake of the crash retrieval and the appearance of men in black in town, said over the air: "We're still trying to figure out what happened on May 14, and we're looking for any information from anyone who saw the event," Hayes announced.[4] He said that on the morning of May 14, right after folks saw the strange object descend, they began calling in reports to 911 and to the radio station. One caller reported to Hayes that "it was coming down fast, but it angled." The object looked like a huge meteorite, but there were no explosions when it hit the ground. David Hayes explained that even though folks might have thought it was a meteorite at first, when it came down from the sky, a team of helicopters retrieved it. In fact, Hayes reiterated, because the aerial recovery team got to the site so fast, in a matter of minutes, he believed they were tracking the object as it fell. "You couldn't have gotten there so quickly unless you were staged and ready to go," he argued. He also suggested that some folks saw the actual impact, as Riverboat Bob did, albeit from a distance alongside the bank. But what struck him personally was the intense presence of military units in the town.[5]

Hayes stated that he saw some unmarked official looking vehicles coming off the freeway. But one was really strange because it was sporting multiple antennas and a dome atop its cabin. And its front license plate read US GOVERNMENT and it had blacked out windows and a guy inside who was staring at out at passersby. Later, Hayes saw the same vehicles outside the radio station, watching. He described the scene: "It was actually parked about 100 yards from the station. And the kinds of equipment and the vehicles they were in was like high-tech possible surveillance."[6] But surveillance on the ground from unmarked vehicles was only one of the strange events after the crash. Folks also remarked about unfamiliar conventional aircraft flying over the city, helicopters without any marking but with black paint instead of olive.

At the time of the appearance of the men in black, local residents near the local airport at Bullhead, Arizona, right across the border with Nevada and California, saw white passenger aircraft with a red stripe along the fuselage. Although otherwise unidentified, to some residents in the area these planes were very familiar. The planes, David Hayes said, were Janet Airline flights, a government-owned CIA airline that ferries scientists and other VIPs back and forth to Area 51 from the Las Vegas McCarran airport. Why were these planes being seen at the airport serving Needles the day after the crash of the object along the Colorado River? Eyewitnesses had reported that on the evening of May 14, they saw a Janet Airlines plane land at Bullhead City airport under the cover of darkness.

Perhaps the best eyewitness to the crash was Frank Costigan, friends with David Hayes, who had had professional experience with observing and identifying controlled flights. He knew what to look for and what was anomalous. Hayes was specific in describing the details of his observation, noting that "all of a sudden the ground seemed to light right up, not like the sun or the moon, but enough to make me look up."[7] The object he said he saw was round in front with flames coming off it. It was blue and turquoise and "five or six times longer than it was wide." He said that he thought he was about to hear a sonic boom when the object came closer or the sound of a crash as it descended over a ridge line near the river. But, he said, he "didn't hear a thing."

From what he could see of the object and its trajectory, he said that it couldn't be anything but out of the ordinary. He said that the object was nothing close to what he had seen before, nothing like a commercial or private aircraft or similar to anything he had seen in the military. "It had to be something like a controlled flight of some sort." Frank said. "That maybe went haywire." That was his professional opinion.

Frank's description of a craft that had been a controlled flight immediately raised three questions: whose was it, who's controlling it, and where did it come from? Local residents knew that the desert around Needles was ringed by military bases and facilities. Just to the northwest was the China Lake Naval Range, the marines' Twenty-Nine Palms and the army's Fort Irwin were to the southeast, further to the east was the Yuma Proving Grounds where the army tested long range artillery, and to the north of Needles was Area 51. In its basic terms, Needles sat in a military hotspot where, for decades, strange craft of all kinds were tested. This opened up question of origin: theirs or ours, an alien vehicle, or one of our pieces of controlled flight ordnance, such as a small cruise missile. Frank also said that entire Mojave desert area was surrounded by military test facilities and that he wouldn't be surprised if the object was meant to land in nearby Lake Mojave but missed its mark and came down along the Colorado River instead, which would also explain why so many military helicopters seemed ready to retrieve this object within fifteen minutes of its impact.

Just like the description given by Frank Costigan, Riverboat Bob's description was detailed and precise. He said, "As I was sitting on the boat fishing at three in the morning, when the whole front of the boat lit up. I looked up and saw a turquoise blue something falling out of the sky. It came straight across from one side to the other. And it hit. It just made kind of made a thump." He said that it made an impact just west of where his boat was sitting when it hit the riverbank about 200 yards away between the river and the railroad tracks that ran alongside the bank.[8] Within fifteen minutes helicopters appeared. Them a "big one came overhead. That thing circled over my boat, went to the site, dropped out of sight for a moment, and he next thing I knew it was up and flying. Not only was it well lit by the other helicopters, they were shining their lights on it." Bob could see clearly that the object was still very hot.

The story in Needles would have faded away over the years, just like many other UFO stories that broke through the surface tension of the media's Overton Window, only to disappear in the absence of any commentary from the government. But contrary to the normal course of events regarding UFOs, the national media coverage of the paramilitary unit from Area 51 created so much scrutiny that in 2009, the first year of Barack Obama's presidency, men in black appeared again in town and the black helicopters continued to surveil the area as if they were still looking for something. If the paramilitary group that came to Needles had been looking for any loose nuclear warhead in 2008, then they might well have been still looking for it in 2009, President Obama's first UFO challenge. Maybe that's why his briefers told him he shouldn't be talking about UFOs to Jimmy Kimmel.

The Bucks County UFO Flap

The other UFO flap that welcomed Barack Obama to the Oval Office was the dramatic series of incidents over Bucks County, Pennsylvania, and Hunterdon County, New Jersey, the aftermath of which resonated throughout 2009, with local residents revealing to a UFO conference of witnesses at Bucks County Community College in Newtown that they had seen multiple formations of lights in the sky. The lights appeared over the Philadelphia International Airport, whose tower would not confirm anything anomalous, over the Delaware River bridge crossings, and into the heart of Philadelphia's surrounding suburban and rural counties.

The most compelling stories coming out of the incident involved two separate, but related, sightings in the suburban Philly community of Levittown, a subdivision of single-family homes built after World War II through the 1950s and '60s. At 1 a.m., on June 25, 2008, a resident looking up at the night sky noticed a very large boomerang-shaped object as it crossed over his property. It was moving slowly at about 1000 feet in the air and sporting brightly colored lights at the rear and along the sides. The object seemed to emit a low, rumbling vibration, not at all like a jet but like something that was pushing air past it. And it traversed the sky overhead for about five minutes before disappearing into the distance. The witness reported he was stunned at the sight.

In another Levittown incident, this one presenting the Pennsylvania state MUFON investigators with hard physical evidence of its presence and its effect on living things, at a little after four in the morning, a local resident in that same subdivision over a week later on July 8, 2008, not only had a sighting, but had a "close encounter of the third kind" when she got up to let her dog out into the backyard. The witness had already seen strange lights in the sky weeks earlier, but on this morning, she saw the same lights again, but much closer, which allowed her to make out the shape of a large floating boomerang overhead.[9]

The craft was moving slowly, prompting the witness to think that it was looking for something or scanning the ground, but, she told the lead investigators John Ventre and Bob Gardner at Pennsylvania MUFON, the object seemed to be "skipping" through air as if it were making stutter steps over her backyard. UFO historians remember that in the latter years of the first Truman administration, right before the Roswell story broke, private pilot Kenneth Arnold, flying a scouting mission over Mount Rainier in Washington, also saw an echelon formation of crescent shaped objects skipping through the air just like, in Kenneth Arnold's words, a saucer skipping across the surface of water. Hence, the term "flying saucer" came into being as a newspaper description of the UFOs.

The Levittown witness also noted that the object seemed to emit a "blue fog" from its tail end which was showering the area with sparkles as if they were tiny, but very bright, silver flakes, "falling down into the treetops and filtering down through the trees, but hovering three or four feet off the ground.[10] The flakes enveloped a tree in the witness's backyard, making the entire tree glow silver as if the objects were swirling around it like a mini dust devil. Then, as the beam reversed itself, the object sucked all the flakes back up into itself along with the beam and moved away. The other aspect that should resonate with UFO researchers is that the witness thought she was only watching the craft for about a half hour, forty minutes at the most, but when she went back inside to check the clock, she had been out there for over an hour. What happened to the time?

Some MUFON investigators had a theory based upon more evidence from the Levittown scene as well as from other UFO encounters involving a lapse of time for which the witnesses could not account. The specific evidence they found when they canvassed the witness's property for clues, just as any CSI team would have done, was even stranger than the witness's lapse of time even as it seemed to suggest the reason behind it. The investigators found two astounding things. The first was that the tree, half of which was enveloped in silver flakes within the center of the craft's beam was dried out and dead on one side, but on the side where the beam missed, the tree was young, moist, bright green, and very healthy. What could have aged the tree in a matter of minutes?

The second piece of evidence was even stranger. On the day previous to the UFO sighting, the witness noticed that a baby robin, simply a hatchling, was sitting snugly in its nest in the tree that would be scanned by the craft the following morning. She remembered the little robin, a portent of spring after a hard winter. Yet when she looked on the ground beside the tree on the morning after her encounter, the robin, which she believed was the same one, was lying on the ground aged, shriveled, and dead. What killed it? How could it have been a hatchling one day and dead from old age the next? Makes no sense, unless.

Investigators who've researched other close encounter type cases have remarked over and over again how the presence of something about a close-by craft affects everything from combustion engines to electrical or hydraulic devices to even hard-wired secure computer connections. Bob Lazar, who said he worked at Area 51 and revealed some of the secrets of alien craft propulsion, described a moment at the classified facility when he was shown a candle that was flickering and then seemed to freeze motionless in the air. What made it stop flickering? The flame was still there, but it wasn't moving. A scientist accompanying Lazar down the corridor explained that the candle was still burning, but a test of the propulsion system simply affected the passage of time by stopping it. Hence, Lazar was looking through a window into a room where something made time appear to stop. Other witnesses, especially a police officer from northern Ohio, described a very similar phenomenon when a UFO passed over his car and the engine went dead along with his unit's radio. Yet, when the UFO had passed, the engine came back to life. What restarted the engine? Or, better, was the engine simply trapped in frozen time between the electrical charges from its distributor, thus stopping the engine without shutting it off?

Perhaps this distortion of time, a form of navigation through space-time, not only explains the phenomenon that Lazar saw and police reported, but it explains what happened to the tree and the robin on the witness's Levittown property. This was a clue of importance certainly to the outgoing President Bush and most likely to the incoming President Obama, who would tell Jimmy Kimmel that he wasn't allowed to talk about things like this.

Chapter 20

HILLARY CLINTON'S PROMISE OF ET DISCLOSURE AND THE JOHN PODESTA EMAIL FLAP

J ust like UFO witness and then-candidate Jimmy Carter on the trail in 1976, former Secretary of State Hillary Clinton told supporters and a newspaper reporter in New Hampshire that she would release our country's UFO files as long as such a release would not compromise national security. Indeed, from a two-term United States senator, First Lady during a major UFO flap in the American southwest, and then a secretary of state likely privy to some of our nation's most closely guarded secrets, Hillary Clinton's promise stirred the poignantly fragile aspirations of the ever-frustrated, long-suffering, always-disappointed, disclosure-obsessed UFO researchers, who, across chat rooms and on late-night conspiracy radio talk shows and podcasts, and through frantic private emails, expressed the slimmest of slim hopes that, finally, this day, their day, had come at last. Insofar as UFOs were concerned, they were with her.

The history of Hillary Clinton's peripheral fascination with UFOs goes back to her husband's presidency, of course, when, in March 1997, she was a witness to and might well have been involved in conversations about the Phoenix Lights and the aftermath of dealing with Governor Fife Symington's attempts to dismiss local area residents' concerns that a strange craft was floating over the city, only hundreds of feet above the balconies and terraces in Paradise Valley. We know that Hillary Clinton had visited with Laurence Rockefeller at his ranch in 1995, right before the end of her husband's first term, and we know, know for a fact because she was so captured on camera, that she read at least one major UFO book that she was clutching when she and Rockefeller were photographed, the just-published Paul Davies's *Are We Alone?: Philosophical Implications of the Discovery of Extraterrestrial Life*, a speculative exploration of what it might mean to human civilization were we to discover that we were not alone in the universe.[1]

Of course, over twenty years later, as the Cassini probe flies through Saturn's moons and has discovered what astrophysicists and other scientists believe are liquid oceans beneath the surface of Saturn's sixth largest moon Encledus, the prospect of extraterrestrial life in our own solar system has become much less remote. Moreover, the discovery of a system of exoplanets in the habitable zone of a red dwarf star forty light-years away also raises the possibility that the conditions that fostered life on Earth may well be fostering life in neighboring systems, and not only in our own solar system.

Then, as if she had dropped a conspiracy bomb, in January 2016 candidate Clinton announced that, if elected, she would reveal what type of activities were taking place at Nevada's Area 51. Presumably, having lived steps away from the Oval Office, she would have known about whatever testing and engineering was going on at the secret facility, perhaps even the government attempts to reverse engineer a flying saucer into one of our own weapons. That 1989 story of attempts to reverse engineer alien spacecraft at Area 51 and test flights of US manufactured flying saucers over the Nevada desert was first revealed by self-described former Area 51 engineer Bob Lazar on KLAS-TV in Las Vegas. Lazar, befriended by test pilot John Lear and KLAS and Coast to Coast AM radio host George Knapp, brought his friends out into the desert at night to witness the test flights. Lazar himself told his story on the History Channel's *UFO Hunters*.[2]

Might Clinton have also known about the development of a teleportation device at Area 51? This was a story that a test pilot in 2007 described about an experiment in which he had been asked to operate what was described to him as a type of "flight simulator" only to find out that as soon as the experiment began, and with no sense of motion or the passage of time, he looked around to see that he was in another "flight simulator" at the other end of a large hangar hundreds of feet away from the original device he had just entered. His teleportation from one booth to another was instant. This is a device under the control of our military and intelligence services. What else do we have of this nature, and is Hillary Clinton a party to those secrets?

When folks on the 2016 campaign trail in January 2016 asked her about the UFO secrets she'd learned and what she might disclose from our government's UFO files, she responded, "I am going to get to the bottom of it."[3] Then in March 2016, her campaign advisor and President Clinton's former chief of staff John Podesta stepped up to confirm that Hillary Clinton would do just what she said. Coming from Podesta, who over ten years earlier had joined journalist and UFO researcher Leslie Kean in a Freedom of Information lawsuit to pry loose what they believed were NASA's UFO files surrounding the 1965 Kecksburg crash, this assurance about Hillary's intentions caused excitement in the UFO community.

Kean and Podesta had prevailed in their lawsuit and managed to disgorge some records from NASA, but the records did not reveal any UFO-related information. This, as former British Ministry of Defence analyst Nick Pope has suggested, can be explained away with the realization that one way to hide records from the public is either never to keep them in the first place or to destroy them after they've been reviewed, a form of UFO burn notice.

John Podesta's experience with the Clintons during the Clinton administration, during which Hillary met with Laurence Rockefeller, whose family foundation not only supplied the funds for the Halloween 1938 Mercury Theatre on the Air "War of the Worlds" broadcast, and later assembled his Pocantico Hills Conference to debunk the Condon Report.[4] Her experience with what Rockefeller knew and brought forward regarding the details of UFO encounters that are often overlooked or underreported by the media and tucked away in secret "special access" files by the government has clearly shaped her thinking about what is kept from the public about UFOs.

Hillary Clinton added that she wants to get to the bottom of the UFO question under the specific qualification that whatever details her administration might release must not compromise in any way classified national security information, including the methods and procedures of gathering that information. Hillary would also likely have been privy to any information provided to President Clinton's science advisor, D. John Gibbons, back in 1993 by Laurence Rockefeller, whose conversations with the scientist resulted in the air force's reviewing its story concerning the crash at Roswell. It didn't change the air force's explanation of the event until 1997 when the official story was switched from a weather balloon to a Project Mogul balloon to sniff out the remains of Soviet nuclear testing lingering in the upper atmosphere and crash dummies that Army Air Force officers mistook for extraterrestrials. Secretary Clinton likely would have known all of this, especially after she visited Rockefeller at his ranch, possibly in an effort to forestall his threat to go public with advertisements in the news media about the government cover-up of UFOs.

Rockefeller revealed that during the Clintons' stay at his ranch, he briefed the first couple extensively on what he knew about UFOs and the government's interaction with them over the years. Although, according to the aforementioned article in the *Express*, no one knows for sure except Bill and Hillary Clinton what Rockefeller actually told them, we can speculate that based on the information Rockefeller gleaned from talking to presenters about the science behind UFO cases it is likelier than not that the Clintons' joint interest was piqued.

However, as Bill Clinton said he found out early in his administration, there was a government within the government that kept the secrets of the UFO files from him. Was this the same special access bureaucracy that turned on JFK when he ordered his military and intelligence services to share UFO files with the Soviets at the beginning of NASA's ventures toward the moon? Was this the same bureaucracy that sabotaged Jimmy Carter's presidency at every turn, even sending emissaries to Tehran to get them to withhold the release of the American hostages until after President Reagan and Vice President Bush were inaugurated? Was Bill Clinton impeached partly because, or even absolutely because, his early attempts to get to the truth of the government's UFO files and the truth behind the JFK assassination? And finally, was Hillary Clinton's 2016 presidential campaign sabotaged, both from within the United States secret government, particularly the Defense Intelligence Agency, and from Vlad the Emailer's Russian hackers and Julian Assange's Wikileaks precisely because of her interest in revealing the truth about the US involvement with UFOs? Representatives Jason Chaffetz and Trey Gowdy can complain all they want to about Hillary's emails, but, as one recent tweet concisely articulated, rather than the emails, it was the white males who doomed her campaign.

Unless, of course, it was the secret government protecting the ETs from disclosure.

During the latter years of Bill Clinton's tenure in office, Hillary actually worked with Laurence Rockefeller to draft an argument for the release of UFO information that, UFO historians have said, she intended to present to her husband partly to forestall Rockefeller's threat to reveal the information he had at his disposal about UFOs. In fact, Hillary went so far as to team up with Rockefeller to pressure the president to resist the pushback from what is now being called the "deep government" or the "shadow government" and release the UFO files once and for all, regardless of what may be in them that does not compromise our national security.

Our suspicion is that after talking extensively to Laurence Rockefeller at his ranch in 1995, the Clintons learned some very disturbing things about the UFO secret, so disturbing, in fact, that Bill Clinton literally went to the other side regarding disclosure even though Hillary was a disclosure supporter. What could that secret have been?

One clue about the seriousness of the secret came to us from a one very unlikely source while another clue came via a very expected source, who shall go nameless. This source said that the ETs are already here, they've been here, and, in the most simple of terms, they run things. In fact, after a fashion, they're us.

Also suggesting that ETs have been here on Earth and are actually "quarantining" us within a very restricted zone around our planet is UFO historian and nuclear physicist Stanton Friedman, who recently told an interviewer, "I think they are here. I think they are here to quarantine us, keep us from going out there."[5] Stan Friedman said that he believes that the ETs believe that human beings are "evil" and should be prevented from colonizing other planets.[6]

Our unlikely source, Red Chamberlain, who died many years ago and was actor Mickey Rooney's father-in-law, was in the Army Air Force in the 1940s and thereafter worked at Lockheed Skunk Works close to its president, Kelly Johnson. He told his children that the government will never release the UFO files because all the world's religions would collapse and governments would lose control. However, this sounds very much like the RAND study predicting panic should humans encounter a threat from outer space.[7] Might these fears have so affected Bill Clinton that in 1997 during the Phoenix Lights he went incommunicado and that after Fife Symington did his job by pooh-poohing the whole incident, Bill Clinton pardoned him? What we do know is that Hillary helped Laurence Rockefeller draft a letter to the White House setting forth the best evidence for the existence of UFOs, a letter and a report that, to the best of our knowledge, was never sent.

The unsent letter to the president's science advisor said in part, "Attached are (1) A draft letter to the President which Laurence has been discussing with Mrs. Clinton and her staff; and (2) A draft report on the "best evidence" about UFOs."[8]

While the letter to Bill Clinton from the Laurence Rockefeller group was never sent, in the years since the Clintons left the White House, operatives in their universe have kept on discussing the UFO issue. For example, in 1999, Paul Begala, now a CNN expert commentator, wrote to John Podesta, who would run the Hillary Clinton campaign in 2016, telling him about Internet megamogul Joe Firmage, who decided to create a type of think tank study group specifically to research, among other things, UFOs.[9] While the letter is intriguing and no doubt Paul Begala might spin his way out of being an advocate for UFO disclosure, if so asked on CNN, even more intriguing is that it was sent to John Podesta who, by the turn of the century, would be one of the leading advocates for the release of government UFO-related information.

In the years between her stint as First Lady and her recent 2016 presidential campaign, Hillary was elected to the US Senate from New York twice and served four years as President Obama's secretary of state. In that capacity, she might well have learned a lot more about UFOs. She would have come into contact with the edges of the UFO secret, not only from our Department of Defense, but from the defense agencies of our allies. Stories abound from the Vietnam War era about NSA intercept flights that spotted UFOs over Vietnam and radioed that information to North Vietnamese, Chinese, and Russian antiaircraft emplacements, who were in an agreement with the United States not to shoot at UFOs. Why, were they our common threat? Hillary Clinton would also have come into contact during the first Clinton administration with stories from the UK, particularly from 1995, when a large triangle floated over Cosford, England, and made its way towards Cornwall. It was tracked by radar, observed by police and

RAF personnel, and was also observed by one of the military's climate officers before disappearing.

In the wake of that Cosford incident, the British Ministry of Defence
asked President Clinton's United States Department of Defense whether
they were testing an advanced secret weapon in the skies over England.
The US DoD responded by asking the UK MoD whether they were testing
a weapon and needed the US to comment. Neither side admitted to flight
testing a top secret aircraft. Because this event took place during the Clinton administration, UFO researchers wondered whether the president or
the First Lady had been looped into the communications chain, which the
White House should have been, given the defense department communications between NATO allies.

The Podesta Emails

"The American people can handle the truth about UFOs," CNN's Eli Watkins quoted John Podesta as having said.[10]

"The US government could do a much better job in answering the quite
legitimate questions that people have about what's going on with unidentified aerial phenomena," Podesta said to CNN's Jake Tapper, indicating that
he knew that there was much more to the reality of UFOs than what super-
market shoppers might find in the checkout lines.[11] And the subtext of his
statement was that after his years in the Clinton and Obama administrations, he knew more than he was allowed to reveal publicly.

Eli Watkins also commented that "In regard to Area 51, Podesta echoed
Clinton's call, saying, 'What I've talked to the secretary about, and what
she's said now in public, is that if she's elected president, when she gets
into office, she'll ask for as many records as the United States federal government has to be declassified, and I think that's a commitment that she
intends to keep and that I intend to hold her to.'"[12] And Watkins reminded
his readers that Podesta affirmed President Bill Clinton's attempts to pry
the UFO files loose from the relevant government agencies, saying that he
once "asked for some information about some of these things, and in particular, some information about what was going on at Area 51."[13] But rather
than speculate for himself about the possibility of life elsewhere in the universe, Podesta said that this is the type of information about which the

public will make its own judgment once the relevant UFO files are released. Even during his stint in the Obama administration, contrary to what the president said to Jimmy Kimmel about whether or not he received a UFO briefing, Podesta tweeted, "Finally, my biggest failure of 2014: Once again not securing the #disclosure of the UFO files. #thetruthisstilloutthere cc: @ NYTimesDowd."

The interest among UFO historians was so intense during the 2016 campaign that after Vlad the Emailer's Russian hackers sent their trove of stolen Podesta emails to Vlad's doppelganger Julian Assange, the search for anything resembling a credible UFO disclosure from the person who served both Clintons and an Obama reached a fever pitch. Disclosure sites in the US and the UK combed through every sentence looking for the faintest clue that Podesta was in the know and was ready to reveal all that he knew, or at least that he would encourage Hillary, should she have prevailed in the presidential election, to reveal whatever she could.

However, there was no release of UFO information and the disclosure advocates now bide their time, hoping that the new president might find an angle on UFO disclosure that would inure to our collective benefit.

Chapter 21

THE AGE OF TRUMP: ALIENS, NEW DISCLOSURES, AND CLIMATIC ARMAGEDDON

More Purported MJ-12 Documents

In late spring of 2017, *Midnight in the Desert* radio host Heather Wade received a cache of documents from an undisclosed source, a person describing himself as former military, who said these were documents concerning MJ-12 revelations from as recently as the 1980s describing what the government learned from the EBE—extraterrestrial biological entity, albeit humanlike—who survived the reported March 25, 1948 UFO crash in Aztec, New Mexico. The incident, reported by author Frank Scully in his book *Behind the Flying Saucers,* told of the discovery and recovery of an alien spacecraft that had crashed outside of Aztec near Los Alamos.[1] Though first debunked and then undebunked, the story of the Aztec crash has remained part of the history of UFOs in America for seventy years, comparable to the story of the Roswell crash in 1947.

The newly revealed MJ-12 documents describe how the crash recovery team discovered the bodies of the alien pilots inside the wreck and then, to their astonishment, discovered that humanlike ETs were stored in a cryogenic state onboard the craft. When at least one of the ETs was revived, it explained that its planet had been responding to a message sent into outer space by Nikola Tesla at the end of Earth's nineteenth century but that their return messages to Tesla were not responded to and, thus, they launched missions to Earth.

The ET, who spoke perfect English, referred to a "treaty" ratified during the Eisenhower administration between the president and the ETs, which allowed them to visit American space and interact in a non-hostile way with human beings; such interactions had been ongoing for at least forty years. Part of the interview focused on the nature of the different planetary societies, particularly the advanced nature of the ETs, who said that they had advanced beyond war and needed to warn humans on Earth that they were destroying the planet.

The papers also referred to different types of alien species that were visiting Earth, among them the humanoids, pure energy beings, hostile insect-type beings, and the Greys, robot-like creatures specifically manufactured for long-term space travel and programmed to carry out specific missions.

Although of interest, because the information contained in the newly released MJ-12 documents could have been ascertained from any number of existing books, including Stan Friedman's *Top Secret/Majic: Operation Majestic-12 and the United States Government's UFO Cover-up, The Day After Roswell, Behind the Flying Saucers,* and Nikola Tesla's own journals, there was no immediate corroboration as to the veracity of the documents. Moreover, the nature of the typeface and the condition of the documents, purportedly from 1989, also would have been computer based rather than simply typed.

Fast Radio Bursts

Are extraterrestrials making contact with us and what will Donald Trump do about it? This might not be as academic a question as it seems because very recently there has been an uptick in what scientists call "fast radio bursts" or FSBs, seemingly anomalous bursts of energy from different points far outside our solar system yet apparently from the same sector of space. Could they have, physicists at the Harvard-Smithsonian Center for Astrophysics say, an artificial intelligent origin rather than being the result of a natural phenomenon?[2]

Scientists have picked up fast radio bursts before, but they have been few and far between and have only lasted for less than a few milliseconds. However, late in 2016, "researchers detected what appears to be the first repeating FRB—eleven high-energy radio bursts all coming from a single source, way out in the distant universe.

"Earlier this year, six more FRBs were detected coming from the same location, and researchers managed to pinpoint their location to a faint dwarf galaxy, more than 3 billion light-years from Earth."[3]

The conventional answer, of course, is that there might well be a perfectly natural cause behind these radio bursts, most likely a black hole pumping out energy along its event horizon as it chews up whatever celestial matter is sucked into its maw. But, tantalizingly, Avi Loeb and scientists at the Harvard-Smithsonian speculate in a paper in *The Astrophysical Journal Letters* that "Fast Radio Bursts are beams set up by extragalactic civilizations to potentially power lightsails," otherwise known as photonic propulsion systems or particles of light, traveling at the speed of light, that can be huge photon collectors.[4] Imagine a propulsion system that needs no fuel other than what is radiated from a star, something about which Bill Nye "the Science Guy" has speculated. In fact, according to a NASA proposal, such a lightsail propulsion system, powered by giant lasers on Earth, could get a space craft to Mars in just three days.[5]

According to NASA's Philip Lubin, the technology for lightsails already exists; we can use it with solar power as well as giant lasers to get unmanned craft to nearby planets and can even send craft to other systems in our galaxy. "There are recent advances that take this from science fiction to science reality," Philip Lubin has said. "There is no known reason why we cannot do this."[6]

Photonic propulsion works by a transfer of the energy from photons to a large reflective sail on a spacecraft. As the amount of light energy continues to pump its energy into the sail, the craft accelerates to 30 percent of the speed of light. Assume, for the sake of speculation, that such a craft, most likely a robotic craft with an artificial intelligence engine to make decisions on its own, learn from its environment, and amalgamate what it learns into its programming, could be constructed either in near earth orbit or even in lunar orbit were we to have established a facility on the moon to house technicians and launch facilities. If NASA had the requisite funding and the support from Congress, it could launch an entire fleet of robotic miniature sensor-heavy craft to the outer planets of our solar system or to the asteroid belt to ascertain whether life exists in some form on Encledus, Titan, Europa, or Ceres, all of which are thought to have subsurface bodies of water. Imagine a US president presiding over a mission to the outer planets the way that JFK launched our space program and directed us to the moon and that Nixon actually presided over the first walk on the lunar surface. President Kennedy will forever be known as the space exploration president because he began it. President Nixon pushed for us to get to the moon to fulfill JFK's goal of landing on the moon by the end of the 1960s, Watergate and resignation notwithstanding. If our forty-fifth president harbors dreams of glory stretching beyond the intricate details of satisfying different voter demographics with his domestic policy to being the president upon whose watch we discovered extraterrestrial life in our own solar system, even former presidential chief advisor Steve Bannon would urge him to reach for the stars.

In 1955, former army general and the Supreme Commander in the Pacific during World War II, Douglas MacArthur, told the *New York Times*, "The nations of the world will have to unite for the next war will be an interplanetary war.

"The nations of Earth must someday make a common front against attack by people from other planets."[7]

Are the fast radio bursts, as the scientists at the Harvard-Smithsonian speculated, an organized array of energy beams powering not just one but a fleet of extraterrestrial spacecraft towards Earth from outside our solar system? And if so, are they coming here and why? Are we so close to an extinction event on planet Earth that those who spread their DNA from galaxy to galaxy across the universe coming back to make sure we don't annihilate the colony of life forms they seeded here, us?

President Reagan repeated this to the United Nations General Assembly. Perhaps, in the tradition of looking into space to conquer the challenge of recharging optimism in a population, the forty-fifth president can be even more transformational.

The Discovery of Extraterrestrial Life in our Solar System

In addition to the possibility of Earth's being bombarded with fast burst radio waves from a distant extraterrestrial civilization, or, as Albert Einstein or Nikola Tesla might argue, a civilization from the distant past in another galaxy whose radio waves have traveled across a vast distance for over a billion years, there's the possibility of ET life closer to home, albeit likely not humanoid as we know it. Discoveries of planetary underground oceans on Encledus, Europa, and Cassini and the discovery of oxygen compounds on comets tearing through our solar system might mean that different forms of organic material have evolved into microbial life colonies in the harsh environments of outer space. Plumes of water, likely from hydrothermal vents on Jovian and Saturnian moons, might have been the environments that spawned life there, in ways similar to deep ocean vents on Earth that have spawned strange life forms. We also speculate that very early Earth was a true water world, spawning various forms of oceanic life that ultimately evolved into amphibious life, reptiles, birds, mammals, and us. If future NASA missions gain the support of the administration to pursue the search for life in our solar system and life is indeed discovered beneath the surface of the outer planet moons, it will constitute another evolutionary moment for humankind, the realization that we are not alone in the universe and that our entire concept of creation must be expanded. The potential of life not only within our own solar system, but exoplanets, earth-like planets elsewhere in our galaxy orbiting their own suns, has created a level of excitement such that even an anonymous broadcaster behind a Guy Fawkes mask has announced, albeit prematurely, that NASA might have just discovered life elsewhere in the universe. What a thrill that would be for any president: Ike and JFK helped launch America's space program, Nixon oversaw our first and subsequent moon landings, and Trump presides over the discovery of life among the outer planets and moons in our solar system. This could mark an historic achievement for the current administration.

The Last Generation

The late evangelical minister Jerry Falwell once prophesied that ours will be the final generation before the end of time. Are we staring into the abyss of the End of Days? Or is his evangelical vision of future's end just one of the different interpretations of history? The Greek historian Herodotus, pre-Christian but still prescient, advised his contemporaries that time itself is circular and history repeats itself. What happened once can happen again and again. Hence, the conflict between Greece and Persia, the West and the East, will happen again. And, inasmuch as ancient Greek democracy inspired the western democratic system of government, it is happening now. Thucydides, writing the history of the Peloponnesian War, also believed that history was not just a recounting of events, but words to the wise that what had gone around will come around again because history did not run in a straight line, it ran in cycles. In Judaism, past, present, and future are one, a singularity, Einstein's block universe in which past, present, and future coexist in a single block of four dimensions, the exemplary aspect of which Moses reveals to the Israelites camped at the borders of Canaan at the end of Deuteronomy in his prophecy of the Blessing and the Curse. Christianity is different insofar as it's predicated on the prophecy that time is presented as a straight line leading to a point after which time has ended, Armageddon and rapture, with the return of the Messiah. Islam also has a belief in the coming of the Twelfth Imam at the end of time.

Given the different theories of history, time, the end of days, and the belief that salvation was only granted to the faithful, the United States at its inception—marked in the very first amendment to the Constitution and in the wake of the religious persecution in England that drove the Puritans to North America—established the basic premise that in the new federal government, Congress was barred from enacting any law with regard to the establishment of religion. The Bill of Rights was later applied to the states under the Fourteenth Amendment. When any state actor, a governing body of any type, gets close to any attempt to legislate regarding religion, that is a live wire and requires the strictest standard of scrutiny from any court. Thus, when government policy drifts toward even a quasi policy regarding religion, courts are more likely than not to look upon it with a very critical and jaundiced eye.

Yet, in the throes of the forty-fifth president's early months in office, a policy emerging from the White House involves a quasi-religious sense of a coming Armageddon, a clash of civilizations at a moment of severe crisis for the United States in what the president's chief advisor believes is its "fourth turning" based on the work of generational theorists William Strauss and Neil Howe, who describe their work in *The Fourth Turning* as a type of American prophecy.[8] The late Strauss and his partner Howe argue that all civilizations go through cycles from great success to complacency to crisis and to a reawakening, during which they are revitalized by an emerging generation. Steve Bannon, who has framed his political arguments on the basis of the prophetic political analyses of Strauss and Howe, has said that he believes America will soon be in a war of survival, announcing in 2015 that "It's war. It's war. Every day, we put up: America's at war, America's at war. We're at war."[9] But if Bannon is looking at a coming clash of civilizations between the West and Islam, guess what: the Bible itself tells us that there will be no clash of civilizations because Isaac and Ishmael, the two sons of Abraham, the father of modern ethical monotheism, came together upon the death of their father to bury him. Rather than a clash of civilizations, the Bible prophesies the amalgamation of civilizations.[10]

Strauss and Howe have traced their cycles of growth, malaise, collapse, and reawakening through the centuries, sometimes actually mischaracterizing events like the Crusades or the collapse of Rome or the Hundred Years' War between the English and French monarchies, which really was more about the rightful heirs to their respective thrones and feudal fiefdoms than anything else. In essence, it was a real-life *Game of Thrones* that actually began at the very end of the twelfth century and gained momentum after Prince John demanded an act of loyalty from his feudal lords by forcing them to choose between their English and French fiefdoms. This was not a collapse of civilization, nor was the fall of Rome, which became the Roman Catholic Church, nor were the Crusades, which resulted in the Renaissance. Evolution, yes; collapse, no. However, the point here is not about the Strauss and Howe interpretation of history per se, but the effect it continues to have on government policy, specifically Bannon's Leninist vision of the end of government itself. This is why that screeching sound you hear every night in the media is Bannon's driving the administration's car against the institutional guardrails of the United States Constitution.

One has only to look at the first five books of the Bible to understand that therein lies the foundation of what this administration wants to destroy, the administrative state. Besides the foundation of much of the western legal system, specifically contract, probate, and criminal law and remedies in tort, the Torah also describes the beginning of public health administration, hierarchical management, and Jethro's articulation of an organized judiciary that would become Article Three of the United States Constitution.

How does an administration gird itself for a clash of civilizations? How does it prepare for an Armageddon? Simply stated, how does an institutional government prepare itself for nothing less than the end of time? Worse, if, as the secret history of the United States and UFOs reveal, what happens when the exoplanetary species that first planted us here decides that it will not allow us to destroy ourselves? As *Real Time's* Bill Maher has asked, does Gene Roddenberry's United Federation of Planets have to intervene? Most people thought—which is what the CIA wanted them to think—that the motion picture *The Day the Earth Stood Still* was pure science fiction. It was science, all right, but most likely not fiction. Nor, by the way, was Isaac Asimov's *Foundation* trilogy, perhaps soon to be a television series coming to a streaming video viewport near you. And if folks at the Harvard-Smithsonian are correct in their speculation and a fleet of light-sailed high photonic energy propelled spacecraft are on their way from a star system beyond our solar system, that is exactly what we might be facing right now.

If the apocalyptic prediction of the end of civilization as we know it, as promulgated in part by Strauss and Howe and advocated as a form of ad hoc policy by the forty-fifth president's senior advisor, is even remotely operative, then the president and his former advisor are like Thelma and Louise driving us toward the edge of oblivion. They are doing this amidst the lynch-mob chants of "lock her up" from those still in fury over the loss of times gone by that will never, ever, come again: natural gas is cheaper than coal, mining companies can blow the tops off coal seam mountains with ten people rather than dig into the mountains with a hundred people, the largest limousine service is Uber who owns no cars, the largest hotel chain is Airbnb who owns no hotels, the large growth in the tech customer service industry has gone overseas, and our American population is aging and not reproducing itself to the point where the per capita productivity of American workers is dropping like rocks off a cliff. That's why "Those Were the Days" is only a theme song sung by Archie Bunker.

We now have statistics that indicate that the we are on the verge of a sixth extinction of life on Earth, and that the demographic that voted this administration into office has the highest rates of suicidal behavior, is suffering from addiction from various kinds of opioid drugs, is suffering amidst its own loneliness, and is boiling over with anger. Add to this recent repeals of environmental protection executive orders, which will most likely accelerate the catastrophic climate change that looms over our future, and you have the recipe for the very collapse of civilization that Steve Bannon has predicted. And speaking of predictions, specifically in response to Trump's pulling the United States out of the Paris climate agreement, not that it was binding upon the United States anyway, physicist Stephen Hawking told the BBC's science correspondent, Pallab Ghosh, that Trump's environmental policies could make climate change irreversible. Specifically, Professor Hawking said, "We are close to the tipping point where global warming becomes irreversible. Trump's action could push the Earth over the brink, to become like Venus, with a temperature of two hundred and fifty degrees, and raining sulphuric acid."[11] As a result, Hawking continued, "I fear evolution has inbuilt greed and aggression to the human genome. There is no sign of conflict lessening, and the development of militarized technology and weapons of mass destruction could make that disastrous. The best hope for the survival of the human race might be independent colonies in space."[12] In other words, for God's sake get out.

Evolution's Path and the End of Humanity as We Know It

Bubbling up now in the early days of the Trump administration is a new proposal by Tesla Motors and SpaceX founder Elon Musk to develop new technologies that will enhance human brain power by linking brains to computer systems. In short, this is an evolutionary moment, taking the human species to a new level of interconnectivity, biologically merging artificial intelligence systems with human beings in a way that will fundamentally alter the human race.[13]

A recent example of this brain/machine interface is a system known as BrainGate2 in which surgeons at Cleveland medical centers implanted two electrode arrays into areas in the paralyzed patient's brain that, before he was paralyzed from a cycling accident, controlled his right arm and hand. They implanted another set of electrodes in the muscles of his right hand and arm, essentially rebuilding the neural connection artificially between his brain and his limb. The electric signals from his brain, generated by the patient's volitional thoughts of moving the paralyzed limb, are translated by a computer into electrical stimuli transmitted to the affected muscles, thus creating movement.[14] This was also reported in the medical journal, *The Lancet*.[15]

Successful integrations of computer circuitry implants in human bodies notwithstanding, Musk's plan is an exponential advance in human/

cybernetic interconnection that throws open a whole new set of issues. Does human computer interconnectivity lend itself to forms of hacking that can implant thoughts and govern actions? Can human beings become deactivated like the "hosts" on television's *Westworld*? Are we looking at a new race of "Six Million Dollar Men"? Or, in a world envisioned by science fiction writer Arthur C. Clark, is this "childhood's end"?

These are some of the challenges facing the Trump administration in the next four years, a version of Moses' blessing and curse prophecy from the end of Deuteronomy. Is the Trump/Bannon mission of deconstructing the administrative state a blessing for those who consider themselves powerless, or a return to the industrial robber-baron oligarchy of the nineteenth and early twentieth centuries? *Quo vadis?*

Climate Change and First Contact

One of the corollary but nevertheless consequential results of climate change and the warming of the glaciers is the thawing from beneath the permafrost of animals whose carcasses were infected millions of years ago with bacteria or strains of virus that have not been seen on this planet since. Accordingly, bacteria that infected animals and even humans hundreds of thousands of years ago might remain in a kind of frozen hibernation ready to be awakened as the permafrost melts due to climate change and global warming and reinfect populations, most likely animals first, causing epidemics for which modern medicine has no cure.[16] In this way, climate change poses a bacterial and viral threat as well as a general environmental threat to the human species, which, some scientists say, is already on the cusp of the next extinction of life on earth, the sixth extinction. This is another challenge facing the forty-fifth president of the United States, especially as the forces within his administration are anti-institutionalist and even anti-government. Unfortunately for those voices, climate change and bacteria sequestered beneath the permafrost for hundreds of thousands of years are completely apolitical, seeking only new hosts to allow them to multiply and thrive, migrate from species to species, until there are no more species left.

The indications that climate change is affecting life on earth are right before our eyes even as pods of usually solitary humpback whales form up in the ocean, as if their whalesong is a good-bye serenade. Other species are experiencing die-offs as well, possibly creating lasting effects on our food chain, effects which might not be felt for generations, but ultimately will be felt. And, climate scientists and weather specialists predict that as a result of rising sea levels, two billion people, that's about one-fifth of the planet's projected population, will be forced from their homes. How far is this from the speeches the candidate made when he promised, in a grand tradition, "un nuevo El Dorado para su Descamisados."[17]

This is not the first time melting ice caps converted land surface to sea bed. According to the writers at *Space Daily*, when the vast Eurasian ice sheet that covered the continent from Britain to Siberia melted and calved away huge portions of itself 23,000 years ago, sea levels rose around the world. The high sea levels lasted for about five hundred years, uniting all the rivers in Europe into a single flow and more than likely forcing the Black and Caspian Seas to flow into each other. This was Noah's flood.

Therefore, one of the most alarming effects of climate change that will have direct effect upon those children alive today is the melting of the Arctic and Antarctic ice caps, the results of which will be a rise in sea levels by the end of this century of between ten and thirty feet. This means, to put it in realistic terms, that if some of the more dire predictions come true and children born within the last decade will have life spans of between eighty and ninety years, Ivanka Trump's children—Donald Trump's grandchildren—will not be able to live in Trump Tower by 2080 because it will likely be flooded, as will Mar-a-Lago and, because it was built on a swamp, the White House. Coastal and low-lying area flooding looks to become a serious consequence over the next seventy-five years if the climate change predictions are correct, a consequence that will have enormous economic impact as well as social dislocation. Among the worst of the aftermaths of Hurricane Sandy, for example, were not only the flooding of New Jersey's Long Beach Island, but the flooding to a state of inoperability of New York City's subway system and power outages that struck large sections of lower Manhattan. Though temporary, these effects were a portent of what could become a regular event not just of severe storms, but of storms and high tides in general. According to the *New York Times*, over 60 percent of Earth's freshwater composes the ice sheets in Antarctica. Just imagine how many of Earth's coastal cities would be underwater if climate change resulted in a significant melt-off of the Antarctic ice cap.[18]

The encroaching sea is already putting at risk Virginia's Tangier Island in the Chesapeake Bay, which, according to a recent story in the *Washington Post*, shrinks by about fifteen feet each year.[19] Despite a blithe promise from President Trump to the village's mayor James Eskridge that everything will turn out OK, the advancing seas are inexorable and only an expensive system of dikes will protect the island from sinking beneath the waters of Chesapeake Bay. Simply put, the disappearance of low-lying islands as a result of the rise in sea levels is not just happening in the South Pacific or the Indian Ocean, but right here on the East Coast of the United States even as Senator James Inhofe throws a snowball into the Senate chamber to demonstrate that temperatures tend to fall below freezing in winter. Hence, no global warming.

Another serious result of flooding took place at the Svalbard Global Seed Vault in Spitsbergen, Norway, the doomsday repository for the world's food crop seeds, preserved there in a deep freeze in the event of a global disaster. As a result of warming in the Arctic Circle, the tunnel to the deeply buried in permafrost vault became flooded. What should have been snowfall in the Arctic summer had turned into rain and the high temperatures melted parts of the permafrost. If the seeds in storage in the vault are lost, humans on planet Earth will have lost our one legacy of civilization that allowed us to evolve from caves into organized agricultural societies. For most of the short-sighted old men of the Senate who most likely won't be alive by the three-quarter mark of this century, the flooding of the seed vault might not seem like a big deal. But to their grandchildren who well might experience crises from the flooding of coastal cities, viral plagues resulting from the release of bacteria long buried in permafrost and now making first contact with human beings, and the loss of scores of thousands of years of human agricultural development, it could mean an extinction event.[20]

Thus, those who may scoff at climate change and now sit behind the wheels of government may laugh at the alarmists or may revel in their own dreams of glory, but there will come a point all too soon when human voices wake them and they will drown.

NOTES

Chapter 1

1. Winthrop, John, *The History of New England from 1630 to 1649* (Boston: Little, Brown and Company, 1853), 349-350.
2. Ibid.
3. http://www.ufoevidence.org/cases/case487.htm)
4. http://www.astrosurf.com/luxorion/ltp-NASA_R-277-1500-1799s.htm and referenced further in Noory, George and William Birnes, *Worker in the Light* (New York: Forge, 2006).

Chapter 2

1. http://www.ushistory.org/VALLEYFORGE/washington/vision.html
2. Ibid.
3. "Wyrd oft nereð unfǽgne eorl þonne his ellen déah."

Chapter 3

1. de Fontenelle, Bernard le Bovier, *Conversations on the Plurality of Worlds* trans. H.A. Hargreaves, ed. Nina Rattner Gelbart (Berkeley: University of California Press, 1990).
2. Voltaire, *Candide* (New York: Bantam Classics, 1984).
3. http://penelope.uchicago.edu/Thayer/E/Journals/TAPS/6/Baton_Rouge_Phenomenon*.html
4. http://www.lpl.arizona.edu/impacteffects/
5. Ibid.

Chapter 5

1. https://www.youtube.com/watch?v=HeSPSm4_-oU

Chapter 6

1. *Filer's Files*, http://nationalufocenter.com/2015/01/filers-files-04-2015-ufo-roosevelts-home/
2. Ibid.
3. Ibid.
4. Ibid.
5. Ibid.
6. Birnes, William J., and Joel Martin, *The Haunting of America* (New York: Forge, 2009), 134-135.

Chapter 7

1. Sanderson, Ivan T., *Invisible Residents* (New York: Avon, 1973) and Jessup, Morris K., *The Expanding Case for UFOs* (New York: Citadel, 1957).
2. Corso, Philip J. and William Birnes, *The Day After Roswell* (New York: Pocket, 1997).
3. Hester, Jan, *UFO Magazine* (Vol. 14.11, November 1999).
4. Private interviews with George Hoover.
5. http://www.presidentialufo.com/franklin-d-roosevelt/60-fdrs-secretary-of-state-and-the-alien-bodies-1939
6. Ibid.
7. https://www.youtube.com/watch?v=EPKGJNk1Qu8
8. https://www.youtube.com/watch?v=9HiCx6Wj8Lk
9. http://www.noufors.com/Franklin_D_Roosevelt.html
10. Ibid.
11. http://sounds.mercurytheatre.info/mercury/381030.mp3
12. http://www.telegraph.co.uk/radio/what-to-listen-to/the-war-of-the-worlds-panic-was-a-myth/
13. http://www.koreatimes.co.kr/www/news/nation/2009/12/113_56715.html
14. Bernstein , Jeremy, *Hitler's Uranium Club: The Secret Recordings at Farm Hall* (New York: Copernicus/Springer, 2001).
15. Dawidoff, Nicholas, *The Catcher was a Spy: The Mysterious Life of Moe Berg* (New York: Vintage, 1995).
16. Cf. Nazi UFOs on *UFO Hunters,* https://www.youtube.com/watch?v=ohgklXi2qvo&list=PL98358E2F3125F98B

Chapter 8

1. https://vault.fbi.gov/UFO/UFO%20Part%201%20of%2016/view
2. Maccabee, Bruce, Ph.D., *The FBI-CIA-UFO Connection: The Hidden UFO Activities of USA Intelligence Agencies* (CreateSpace, 2014).
3. http://www.majesticdocuments.com/pdf/eisenhower_briefing.pdf
4. https://www.youtube.com/watch?v=jAfTY7NuceQ and the Roswell crash debris at the Pentagon in 1961, cf. Corso, Philip J., and William Birnes, *The Day After Roswell* (New York: Pocket, 1997).
5. Marcel, Jesse, Jr., *The Roswell Legacy: The Untold Story of the First Military Officer at the 1947 Crash Site* (Wayne, NJ: New Page Books, 2007).
6. Carey, Thomas and Donald Schmitt, *Witness to Roswell* (Wayne, NJ: Career Press, 2009).
7. Jacobsen, Annie, *Area 51: An Uncensored History of America's Top Secret Military Base* (New York: Little Brown, 2011).
8. Cf. Dwayne Day at http://www.thespacereview.com/article/1852/1
9. http://www.roswellfiles.com/Witnesses/brazel.htm
10. Ines Wilcox, "My Four Years in the County Jail," in "The Children of Roswell," *UFO Magazine,* 13 (7) (November 1998), 51.
11. http://roswellproof.homestead.com/haut.html
12. Birnes, William J. and Mark, Merritt and Natalie Magruder, "Squiggly," *UFO Magazine,* 21 (4) (June 2006) 33-39.

13. Corso, Philip J. and William J. Birnes, *The Day After Roswell* (New York: Pocket, 1997).

14. McCartney, Patrick, "Shulman's Sensation," *UFO Magazine*, 13 (3) (March-April 1998), 9.

15. Sampson, Paul, "Investigation On in Secret After Chase Over Capital [image of *Post* front page from http://www.rense.com/general8/flew.html]," *Washington Post*, July 28, 1952.

16. Feschino, Frank, Jr., *Shoot Them Down* (Self Published, 2007).

17. Ruppelt, Captain Edward J. , *The Report on Unidentified Flying Objects* (New York: Ace, 1956).

18. https://www.youtube.com/watch?v=tcRtkA1Rmvw

19. https://www.youtube.com/watch?v=k8unOAZtjkA

Chapter 9

1. Berlitz , Charles and William Moore, *The Roswell Incident* (New York: Grosset, 1989).

2. http://www.bahaistudies.net/asma/gerald_light.pdf

3. http://www.bibliotecapleyades.net/exopolitica/esp_exopolitics_Q_0.htm

4. Scott, William B., Michael Coumatos and William Birnes, *Space Wars* and *Counter Space* (New York: Tor Books, 2008 and 2009, respectively).

Chapter 10

1. Lertzman, Richard A. and William Birnes, *Dr. Feelgood* (New York: Skyhorse, 2014).

2. Ibid.

3. Burleson, Donald, *UFOs and the Murder of Marilyn Monroe* (Black Mesa Press, Roswell, NM, 3rd ed., 2003).

4. http://www.blackmesapress.com/page4.htm

5. Ibid.

6. http://www.bibliotecapleyades.net/sociopolitica/esp_sociopol_mj12_3k.htm

7. Lertzman and Birnes, *Dr. Feelgood* (New York: Skyhorse, 2014)

8. Cf. http://www.spyculture.com/fbi-report-on-marilyn-monroe-death-conspiracy-theory/

9. Lertzman, Richard A. and William Birnes, *The Life and Times of Mickey Rooney* (New York: Gallery, 2015).

10. Milan, Mike, *The Squad* (New York: Shadow Lawn Press and Carol Publishing, 1989).

11. Ibid.

12. http://www.spyculture.com/fbi-report-on-marilyn-monroe-death-conspiracy-theory/

13. Ibid.

14. Lertzman and Birnes, *Dr. Feelgood* (New York: Skyhorse, 2013).

15. Lertzman, Richard A. and William Birnes, *Dr. Feelgood* (New York: Skyhorse, 2014). See also Kempe, Frederick,. *Berlin 1961* (New York: Penguin, 2011).

16. *Loving v. Virginia*, 388 U.S. 1 (1967).

17. Birnes, William J., *UFO Hunters, Book 1* (New York: Tor, 2011).

18. Cf. Marden, Kathleen, and Stanton Friedman, *Captured! The Betty and Barney Hill UFO Experience* (Franklin Lakes, NJ: New Page, 2007).

19. Birnes, William J., *UFO Hunters Book I* (New York: Tor, 2011).

20. Marden, Kathleen, and Stanton Friedman, *Captured! The Betty and Barney Hill UFO Experience* (Franklin Lakes, NJ: New Page, 2007).

21. Fuller, John G., *Interrupted Journey*, introduction by Dr. Benjamin Simon (New York: Dial, 1966).

22. https://www.youtube.com/watch?v=lpGA8ctZN34&list=PL50960F2A238B05A6 and private email

23. https://www.youtube.com/watch?v=lpGA8ctZN34&list=PL50960F2A238B05A6

24. http://www.dailymail.co.uk/news/article-1378284/Secret-memo-shows-JFK-demanded-UFO-files-10-days-assassination.html

25. Ibid.

26. Ibid.

27. Lertzman and Birnes, *Dr. Feelgood*.

28. http://www.newsweek.com/johnson-tapes-173886

29. Corso, Philip J., private conversation with William Birnes.

Chapter 11

1. "Symposium on Unidentified Flying Objects," Submitted to the House Committee on Science and Astronautics, Rayburn Office Building, Washington, D.C., July 29, 1968.

2. Friedman, Stanton, *Top Secret / Majic* (New York: Da Capo, 1996).

3. https://www.youtube.com/watch?v=RoRRBJC7QRw

4. https://www.youtube.com/watch?v=ouuh10RUgvE

5. https://www.youtube.com/watch?v=DmBsU-FHqa4

6. Druffel, Ann, *Firestorm: Dr. James E. McDonald's Fight for UFO Science* (2nd ed., Taylor, TX: Granite Publishing, 2003).

7. https://www.youtube.com/watch?v=Z3g3EJOZ82s

8. Ibid.

9. Ibid.

10. Ibid.

11. Ibid.

12. https://www.youtube.com/watch?v=3ZzYHBOIN8Q

13. Ibid.

14. https://www.youtube.com/watch?v=MNgrJE6ubOI)

15. Ibid.

16. Ibid.

17. Ibid.

18. Ibid.

19. Ibid.

20. Kean, Leslie, *UFOs: Generals, Pilots, and Government Officials Go On the Record* (New York: Three Rivers Press, 2011).

21. Jim Cohen at http://www.cohenufo.org/SwampGasCase_cpy.htm

22. Ibid.

23. Ibid.

24. https://www.fordlibrarymuseum.gov/library/document/0054/4525586.pdf
25. Salas, Robert, *Faded Giant,* (North Charleston, SC: Book Surge Publishing, 2005).
26. http://www.cufon.org/cufon/malmstrom/malm1.htm
27. Ibid.
28. Ibid.
29. Ibid.
30. Ibid.
31. Ibid.

Chapter 12

1. Sensei, in Japanese martial arts, a skilled teacher with knowledge who has trod the path before his students. Cf. Crichton, Michael, *Rising Sun* (New York: Random House, 1992).
2. Sturrock, Peter A., "An Analysis of the Condon Report on the Colorado UFO Project," *Journal of Scientific Exploration.* 1 (1): 75, 1987, and http://files.ncas. org/condon/.
3. http://badufos.blogspot.com/2014/11/fbi-releases-its-files-on-dr-james-e.html
4. Private interview with Betsy McDonald on the set of *UFO Hunters.*
5. http://badufos.blogspot.com/2014/11/fbi-releases-its-files-on-dr-james-e.html
6. Corso, Philip J. and William J. Birnes, *The Day After Roswell* (New York: Pocket, 1997), 276-332.
7. For a fuller and more detailed coverage of NASA's Apollo transmissions and structures on the lunar surface, see Birnes, William J. and Harold Burt, *Unsolved UFO Mysteries* (New York: Warner Aspect, 2000), 196-222.
8. Cf. Grant Cameron at http://www.presidentialufo.com/old_site/richardm.htm
9. Henry , William A. *The Great One: The Life and Legend of Jackie Gleason* (New York: Doubleday, 1992), 111.
10. Warren, Larry, and Peter Robbins, *Left at East Gate* (New York: Marlow, 1997) and on the Presidential UFO site, http://www.presidentialufo.com/old_site/ richardm.htm.
11. http://www.presidentialufo.com/richard-nixon/87-jackie-gleasons-wife-talks

Chapter 13

1. http://www.presidentialufo.com/old_site/ford.htm
2. http://nationalufocenter.com/category/filers-files/
3. Cf. Norris, Joel, *Serial Killers: A Growing Menace* (New York: Doubleday, 1988).
4. Private conversation with Charles Manson, 1987.
5. http://www.theblackvault.com/documents/ufos/1976iranincident.pdf
6. Ibid.
7. http://www.theblackvault.com/casefiles/the-1976-iran-incident/#

Chapter 14

1. http://www.sacred-texts.com/ufo/carter.htm
2. http://www.forbes.com/sites/jimclash/2017/01/18/mufons-jan-harzan-on-area-51-jimmy-carters-ufo-sighting-and-history-channels-hangar-1-series/#1d261e7956c3
3. Ibid.

4. http://www.democraticunderground.com/discuss/duboard.php?az=view_all& address=104x1956796

5. Ibid.

6. http://www.collective-evolution.com/2016/06/25/pentagon-papers-watergate-lawyer-shares-inside-information-about-ufos/

7. Ibid.

8. Private interview with Seth Shostak.

9. https://www.cia.gov/library/center-for-the-study-of-intelligence/csi-publications /csi-studies/studies/97unclass/ufo.html

10. For an excellent review of Carter's attempts, through both NASA and the UN, to study the question of UFOs, see Grant Cameron's article at http://www .presidentialufo.com/jimmy-carter/93-jimmy-carter-ufo.

11. Cf. Warren, Larry and Peter Robbins, *Left at East Gate* (New York: DaCapo, 1997); Bruni, Georgina, *You Can't Tell the People* (London: Sidgwick & Jackson Ltd, 2000); Pope, Nick, *Encounter in Rendlesham Forest: The Inside Story of the World's Best-Documented UFO Incident* (New York: Thomas Dunne, 2014); Birnes, William, *UFO Hunters, Book I* (New York, Tor Books, 2013), and *UFO Hunters*, episode 105, https://www.youtube.com/watch?v=2tMeAQ7TNuw].

12. https://www.youtube.com/watch?v=2tMeAQ7TNuw

13. Ibid.

14. Ibid

15. Ibid

16. Ibid

17. Pope, Nick, *Open Skies, Closed Minds* (New York: Pocket, 1997) and *Encounter at Rendlesham* (New York: Thomas Dunne, 2014).

18. Bruni, Georgina, *You Can't Tell the People* (London: Sidgwick and Jackson, 2000; rpt. Pan Macmillan, 2001).

19. https://www.washingtonpost.com/politics/how-bannons-navy-service-during-the-iran-hostage-crisis-shaped-his-views/2017/02/09/99f1e58a-e991-11e6-bf6f-301b6b443624_story.html?utm_term=.7f7a4c2b8685&wpisrc=nl_evening&wpmm=1

20. https://www.youtube.com/watch?v=ekuL8IliABc)

21. May Pang and Henry Edwards, *Loving John: The Untold Story* (New York: Warner, 1983).

22. December 7, 2004, http://www.telegraph.co.uk/culture/4730520/The-night-aliens-called-on-Lennon.html

23. http://www.telegraph.co.uk/culture/4730520/The-night-aliens-called-on-Lennon.html

24. *New Yorker,* March 20, 1948, 30

25. Salinger, Jerome David, "Uncle Wiggily in Connecticut," *The New Yorker*, Ibid., and rpt., *Nine Stories*, (New York: Random House, 1953).

26. See Lepore, Jill, "Esmé in Neverland," *The New Yorker* (November 13, 2016), http://www.newyorker.com/magazine/2016/11/21/the-film-jd-salinger-nearly-made/amp

27. Salinger, J. D., *Catcher in the Rye* (New York: Little Brown, 1951).

Chapter 15

1. "All the Presidents' UFOs," *UFO Magazine*, 15 (1) (January 2000), 18-25.
2. http://science.howstuffworks.com/space/aliens-ufos/ronald-reagan-ufo.htm
3. Ibid.
4. http://www.dailymail.co.uk/news/article-2205360/Screen-legend-Shirley-MacLaine-says-Ronald-Reagan-spotted-UFO-1950s—alien-told-switch-careers.html
5. Milan, Michael, *The Squad* (New York: Shadow Lawn Press and Carol Publishing, 1989).
6. https://www.quora.com/Did-William-Casey-CIA-Director-really-say-Well-know-our-disinformation-program-is-complete-when-everything-the-American-public-believes-is-false
7. http://www.bibliotecapleyades.net/sociopolitica/serpo/information27a.htm
8. Ibid.
9. Ibid.
10. Ibid.
11. Ibid.
12. Ibid.
13. Ibid.
14. Ibid.
15. http://www.bibliotecapleyades.net/exopolitica/exopolitics_reagan01.htm
16. Ibid. You can listen to the Vespe interview at http://static3.aintitcool.com/assets2011/SSc7.mp3.
17. Cited in http://www.bibliotecapleyades.net/exopolitica/exopolitics_reagan03.htm
18. Ibid.
19. Ibid.
20. Ibid.
21. Booth, B. J. at http://www.ufocasebook.com/Hudsonvalley.html
22. Ibid.
23. http://www.abovetopsecret.com/forum/thread1101879/pg1
24. Birnes, William, *UFO Hunters Book I* (New York: Tor, 2013), 239.
25. http://www.nicap.org/reports/861117_brumac.8k.com_JAL1628.pdf
26. http://mysteriousuniverse.org/2013/12/pilot-testimonials-captain-kenju-terauchi/
27. Ibid.
28. https://www.youtube.com/watch?v=3NkiO9YxRhw
29. Sanders, James, *The Downing of TWA Flight 800* (New York: Pinnacle, 2013).

Chapter 16

1. http://www.presidentialufo.com/old_site/bush_ufo_story.htm
2. Ibid.
3. Ibid.
4. Ibid.
5. Hopkins, Budd, *Witnessed: The True Story of the Brooklyn Bridge UFO Abductions* (New York: Pocket, 1996).
6. http://www.presidentialufo.com/old_site/bush_ufo_story.htm
7. Birnes, William J., *Aliens in America* (Avon, MA: Adams Media, 2010), 94-105. See also Walters, Ed, *The Gulf Breeze Sightings* (New York: Morrow, 1990).

Chapter 17

1. "All the Presidents' UFOs," *UFO Magazine*, 15 (1) (January, 2000), 23.
2. Ibid.
3. Hubbell, Webster , *Friends in High Places* (New York: Morrow, 1997).
4. https://www.cia.gov/library/center-for-the-study-of-intelligence/csi-publications/csi-studies/studies/97unclass/ufo.html
5. Ibid.
6. Austin , Jon, "What really happened when Hillary and Bill Clinton tried to open UFO truths 21 years ago," *Express* (March 9, 2016).
7. Ibid.
8. Ibid.
9. Sturrock, Peter A., *The UFO Enigma* (New York: Warner, 1999).
10. See articles by Hamilton, William, "The Phoenix Sightings," *UFO Magazine*, 15 (3) (March 2000); Birnes, William, "Arizona Lights," *UFO Hunters Book 2* (New York: Tor, 2015). and History Channel's *UFO Hunters* (https://www.youtube.com/watch?v=LRRrpQyvK5Q&list=PL20D4686A4AF472DC).
11. Austin, Jon, "PHOENIX LIGHTS UFO: New witness comes forward 20 years later . . . and it's a hollywood legend," *Express* (May 24, 2017).
12. https://www.youtube.com/watch?v=Jva43Ujkfqg&list=PL20D4686A4AF472DC&index=2 and https://www.youtube.com/watch?v=bLCHxONo2uM&list=PL20D4686A4AF472DC&index=5_
13. *UFO Magazine*, 22 (3) (March 2007), and https://www.youtube.com/watch?v=HoFj_3iQo2U&index=3&list=PL20D4686A4AF472DC
14. https://www.youtube.com/watch?v=jA8CyOv67zU&index=4&list=PL20D4686A4AF472DC
15. Ibid.
16. https://www.youtube.com/watch?v=Jva43Ujkfqg&index=2&list=PL20D4686A4AF472DC
17. http://www.azcentral.com/story/news/local/phoenix/2015/02/19/phoenix-lights-arizona-ufo-mystery/23677303/
18. https://www.youtube.com/watch?v=Jva43Ujkfqg&index=2&list=PL20D4686A4AF472DC
19. Ibid.
20. Ibid.
21. Ibid.
22. https://www.youtube.com/watch?v=jA8CyOv67zU&index=4&list=PL20D4686A4AF472DC
23. Ibid.
24. https://www.youtube.com/watch?v=bLCHxONo2uM&list=PL20D4686A4AF472DC&index=5
25. Ibid.

Chapter 18

1. http://presidentialufo.com/george-w-bush/183-dick-cheney
2. Ibid.
3. Ibid.